How to Make a Hand-operated Holepunch

Ted Stone and Jim Tanburn of ApT Design and Development

Step-by-step instructions on how to build a versatile machine for punching holes in sheet metal, and for 'nibbling' complex profiles in sheet metal. This manual includes complete details of how to build and use the holepunch and some design options.

Intermediate Technology Publications 1986

Acknowledgements

Financial assistance in the development of the machine and in the production of this manual was made available through the Intermediate Technology Development Group from a grant from the Overseas Development Administration. Technical assistance in the design, building and testing of the machine, and in the production of this manual, was also provided by E.J. Stone. All this assistance is gratefully acknowledged.

Important notice to constructors

The plans and instructions given in this manual *must* be read very carefully at each step of the construction process.

The order described here in which parts are made and assembled is the easiest, and should be followed exactly.

Particular care must be given to the relative positions of parts before they are welded.

Where materials are not available in the sizes specified in the manual, give serious thought as to how your substitution with material of a different size will affect the function of that part of the machine:
— Will the change weaken the finished machine?
— Will it make the machine less durable?
— Will this substitution alter other dimensions given elsewhere in the manual?

Where flat bar or plate of the specified thickness is not available, consider whether you could weld two thinner pieces together around the edges and use this in its place.

It is usually better to use a larger steel section than a smaller one.

Components which slide together or rotate in one another should not be painted on those surfaces, and should be greased or oiled as the machine is assembled. Further oiling from time to time will also prolong the life of the machine.

At the back of this manual is a checklist. Please read it both before you build the machine and after you have completed the machine. If all the points listed are O.K. you will be well pleased with your machine.

9 King Street, London WC2E 8HW, U.K.

ISBN 0 946688 13 3

Printed by the Russell Press Ltd., Bertrand Russell House,
Gamble Street, Nottingham NG7 4ET, U.K.

Contents

Glossary

All dimensions are given in millimetres

∅12	Diameter of 12mm
M12	12mm metric thread
MS	Mild steel
Blank	The piece of metal pushed out when a hole is punched
Tool steel	Medium or high carbon steel e.g. leaf spring

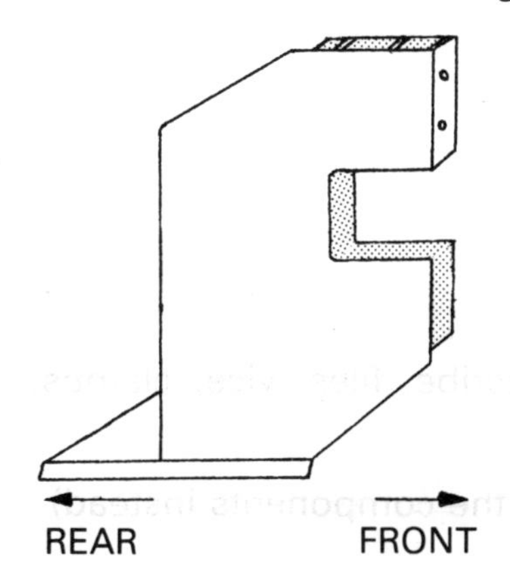

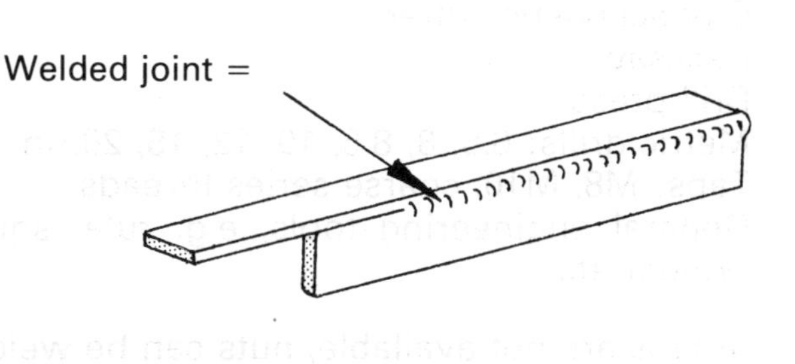

Introduction

This booklet describes how to build and use a hand-operated holepunch. The construction of the machine requires only a small capital outlay and uses steel in readily-available sizes. The tools involved are the simplest possible to make an effective piece of workshop equipment.

Once constructed, the holepunch will make neat holes up to 25mm diameter in sheet steel 1.5mm thick. The punch and die sets can be changed quickly and easily, making a wide range of hole sizes and shapes possible. The throat is adjustable up to a depth of 100mm.

The holepunch will prove most useful in any workshop working with sheetmetal, as it is better in many respects than a drill. Small drill bits break easily and are expensive to replace, while large drill bits are often difficult to use in sheet steel.

The holepunch can be used to make holes for rivets, bolts and screws, to make airholes in stoves and to manufacture washers. It will substantially ease the production of a wide range of agricultural and food processing equipment in many situations where production is currently limited by the tools available.

Finally, a 'nibbler' attachment is also described; this allows the cutting out of very large holes and complex profiles in steel sheet.

Tool and equipment list

The items listed below are needed to build the complete holepunch. Note that the turning, or lathe work, can be given to another workshop if a lathe is not readily available.

Lathe
Electric welder
Oxy-acetylene cutter
Hacksaw
Drill press
Metric drills: 6.5, 8, 8.5, 10, 12, 15, 20mm dia.
Taps: M8, M10, coarse series threads
General engineering tools, e.g. rule, square, scribe, files, vice, clamps, grinder etc.

(If taps are not available, nuts can be welded to the components instead).

Uses of the holepunch

The maximum thickness of the steel which can be punched will depend on the size of the hole; the smaller the hole, the thicker the steel. To give an idea of the capacity, this machine will punch ∅25 holes in 16 gauge steel.

The holepunch can be used in place of a drill for most applications; it will also perform some jobs that cannot be done with a drill. Some of the products which can be manufactured easily with the holepunch are illustrated below.

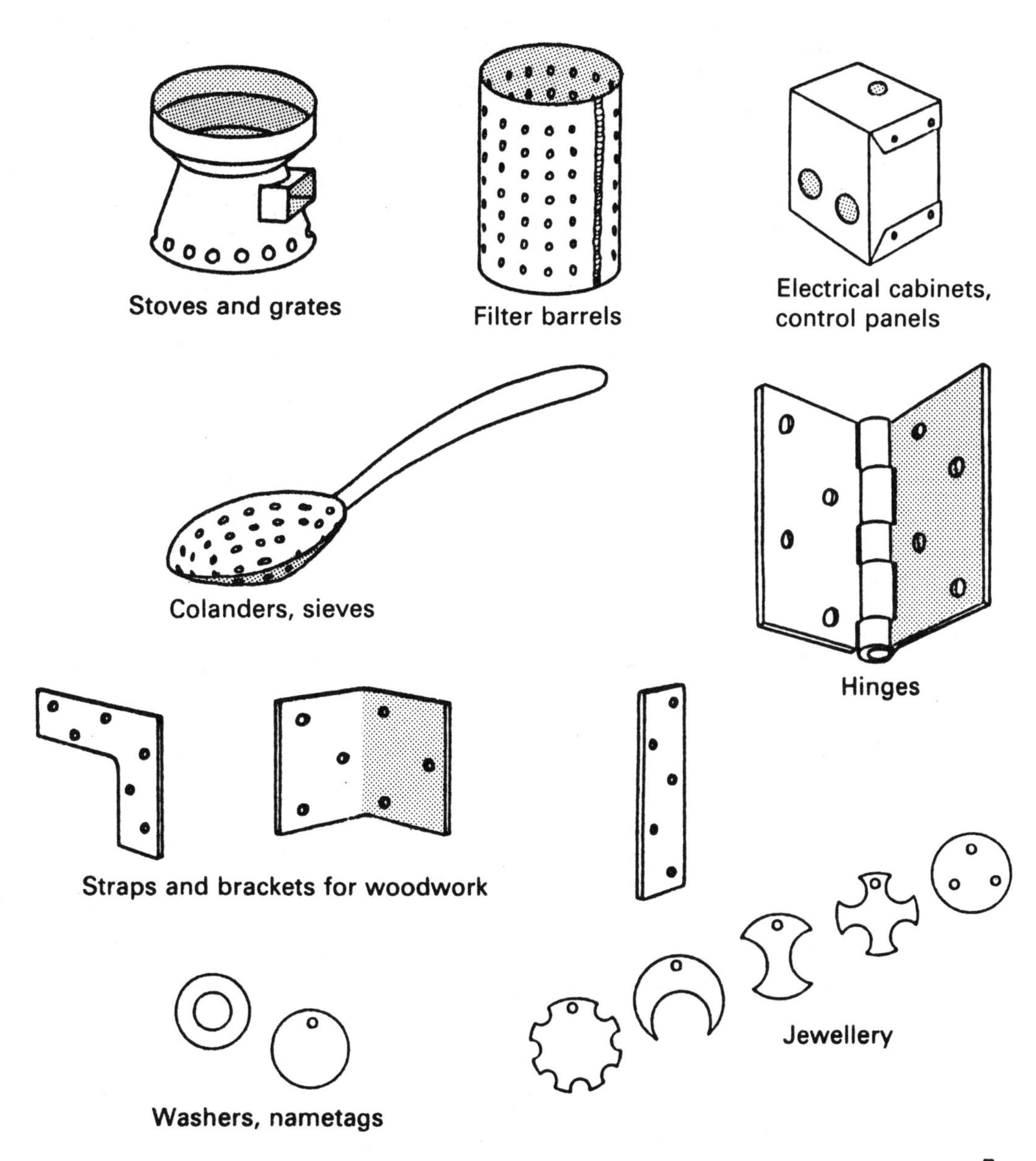

Materials list

Where materials will be cut from a stock length, a kerf of 3mm per item has been allowed for. The kerf is the slit made by the cutting saw. If flame cutting is used then a greater allowance must be made.

Material	Section	Length (mm)
Black flat (mild steel)	40 × 12	3447
	25 × 5	1562
Equal angle (mild steel)	30 × 30 × 5	153
Mild steel plate	8 × 300 × 380	2 (qty.)
Steel tube, medium duty	Nominal bore 25	603
Bright rounds (mild steel)	∅30	263
	∅20	244
High carbon steel		
Scrap half shaft (axle)	∅25	180
Scrap vehicle leaf spring	60 × 12 (approximate size)	180

(Assuming 4 sets with the largest size ∅25 and allowing a 5mm kerf.)

If new material is to be used, then a tool steel or other high carbon steel of more than 0.5% carbon should be used.

Fasteners	Quantity
M12 × 40 Bolts	4
M12 Nuts	4
M12 Flat washers	8
M10 × 40 Bolts (screws)	2
M10 × 25 Bolts (screws)	4
M10 Nuts	4
M10 Flat washers	6
M8 × 30 Bolts (screws)	3
M8 × 25 Bolts (screws)	1
M8 × 20 Bolts (screws)	2
M8 Nuts	10
M8 Flat washers	2
Split pins 3 × 30	4

Description

The holepunch consists of a box-section frame made from two pieces of steel plate in the shape of a 'C'. The lower part supports the die (into which the sheet steel is punched) while the upper part holds the punch operating mechanism.

This mechanism consists of a vertical shaft which holds the punch, and a hand lever which is connected to the shaft by two linkage plates. The guide for the shaft is adjustable, to allow for wear.

When the hand lever is pulled down, the punch is forced through the sheet metal being punched, and the piece pushed out (the blank) drops out through the die.

On lifting the hand lever, the punch is pulled up out of the hole and the sheet metal is held down by the two 'strippers'.

The punch and die are held in place by handscrews, and so can be quickly changed for ones of a different size. There is also a depth gauge which is useful to set the distance of a hole from the edge of the metal being punched, particularly where repetitive work is being done.

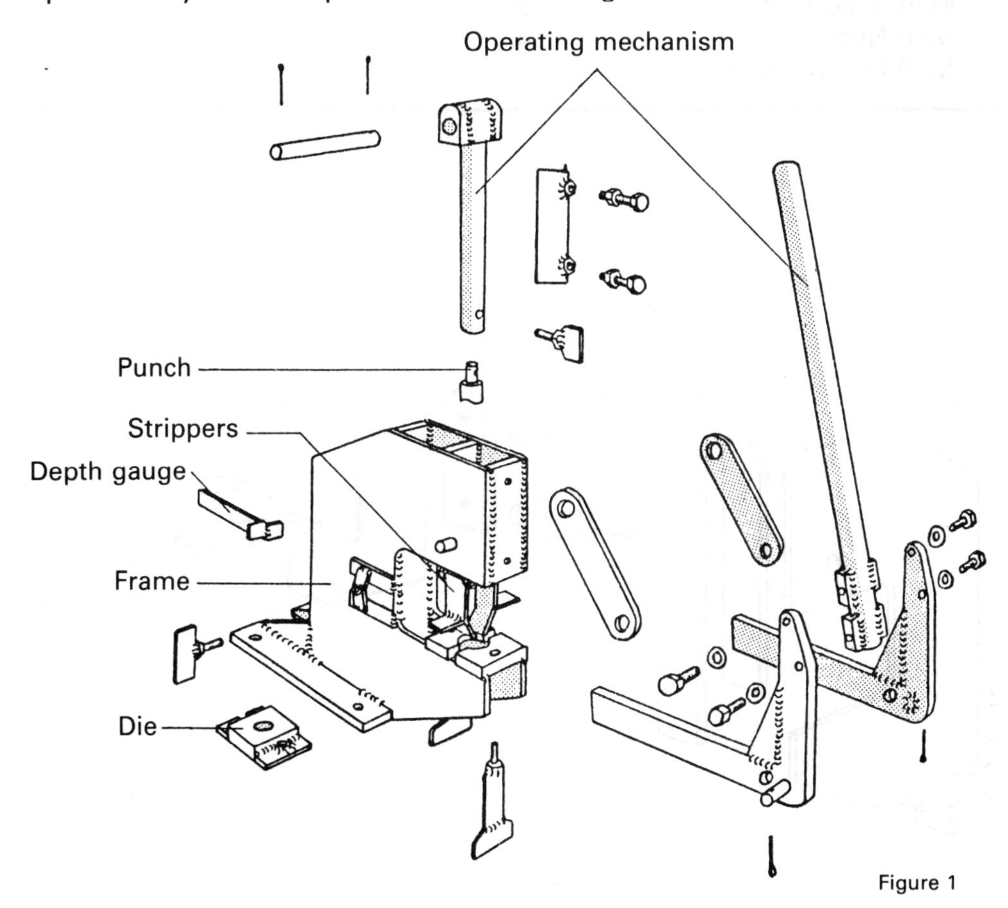

Figure 1

Construction

FRAME ASSEMBLY — A
PARTS

Part	Name	Quantity	Section	Length(mm)
A1	Frame side plates	2	8 MS plate	300 × 380
A2	Cross member (rear)	1	40 × 12 MS flat	306
A3	Cross member	1	40 × 12 MS flat	380
A4	Cross member	1	40 × 12 MS flat	150
A5	Cross member (front)	1	40 × 12 MS flat	150
A6	Fillets	2	40 × 12 MS flat (one piece cut diagonally into two)	95
A7	Bench mounting flanges	2	40 × 12 MS flat	250
A8	Punch holder guide	1	30 × 30 × 5 MS angle	150
A9	Linkage pivot pins	2	∅20 bright round bar	20
Fasteners				
	M10 × 40 Bolts	2		
	M10 Nuts	4		
	M10 Flat washers	2		

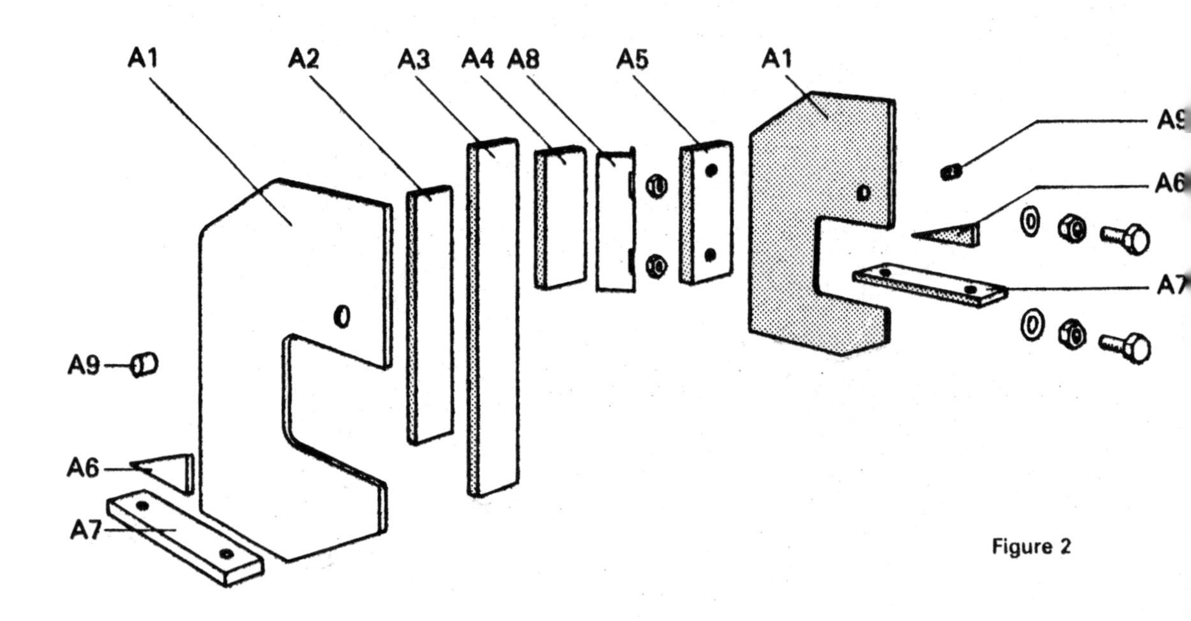

Figure 2

BASE FRAME ASSEMBLY

Cut the steel accurately to the dimensions given in the Parts table for parts **A1** to **A9**. Carry out the additional cutting, drilling and tapping operations required for the components shown in Fig. 3 (tack weld the two pieces **A1** together for final grinding and drilling).

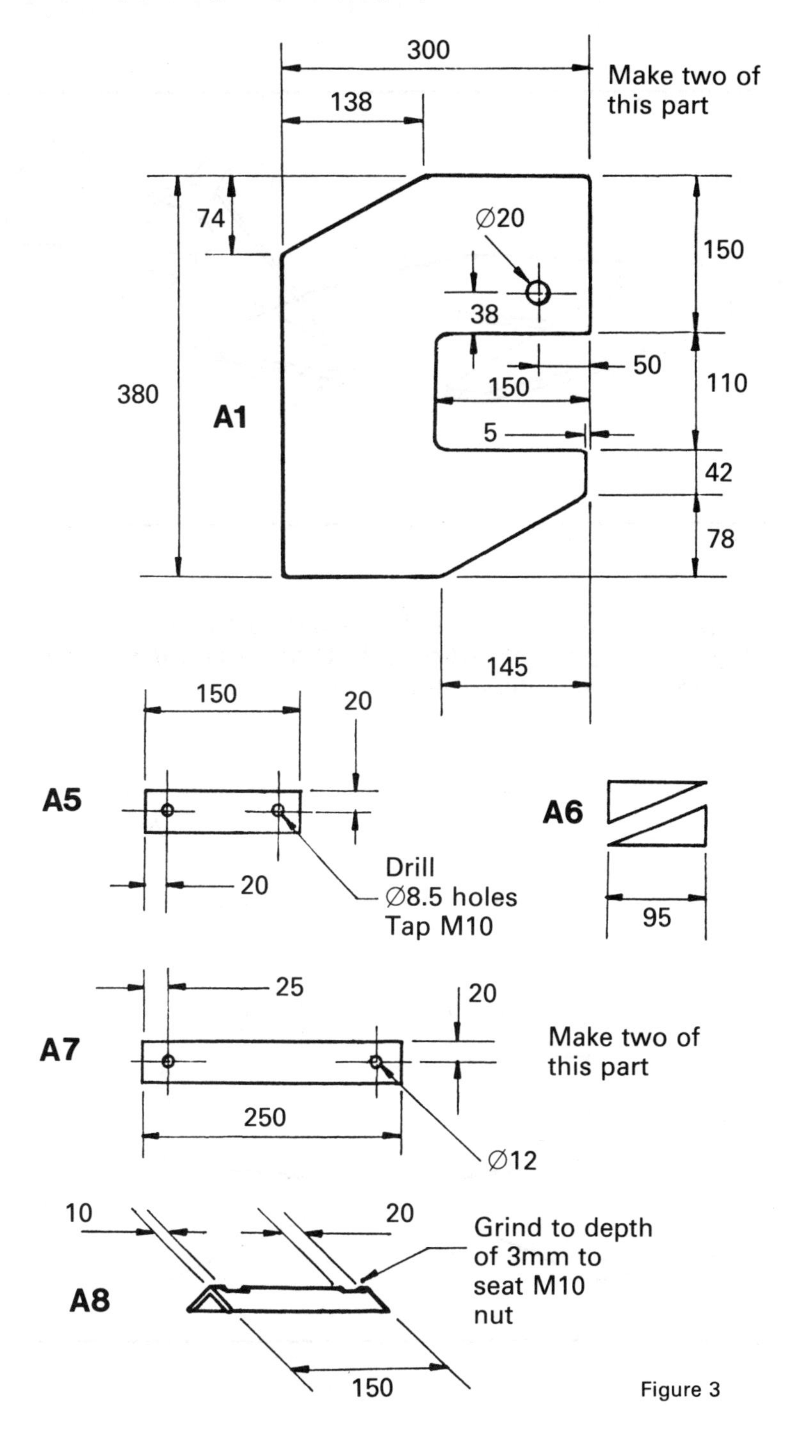

Figure 3

Chamfer (4mm) one end of each of the linkage pivot pins **A9**; place one pin into the hole in one frame side plate **A1**, so that the chamfered end is flush with one side of the plate. Tack weld the chamfered end to the plate (Fig. 4) and check that the pin is square to the plate. Weld in place, allowing the weld to run into the space created by the chamfer. Grind the weld until it is flush with the plate.

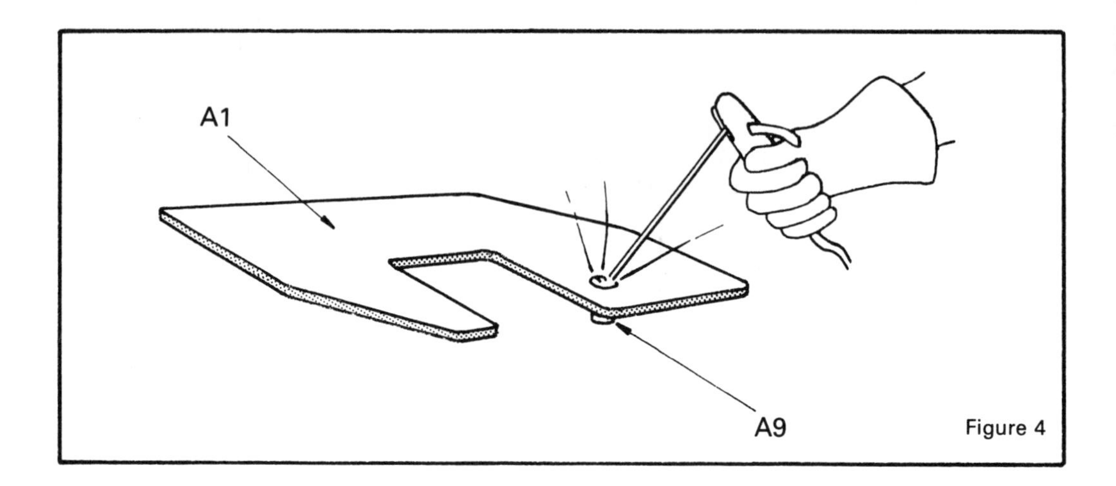

Figure 4

Repeat this operation with the other frame side plate **A1** and linkage pivot pin **A9**, ensuring that the pin is fitted on **THE OPPOSITE SIDE** of the plate to the one already fitted. This is illustrated in Fig. 5.

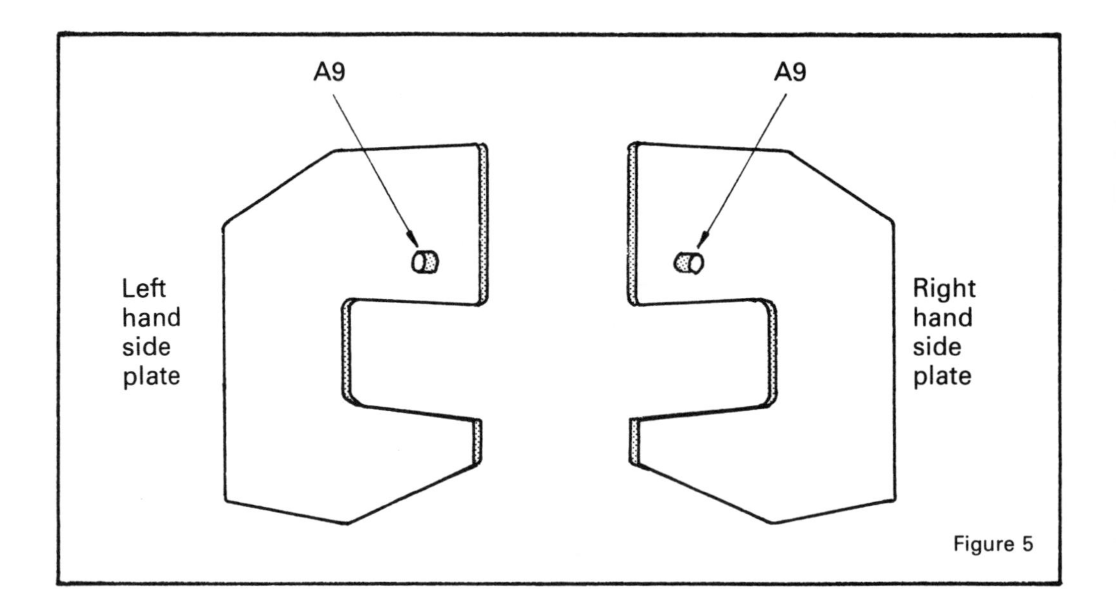

Figure 5

Drill out two of the M10 nuts to ∅10. Screw the two M10 bolts through the cross member **A5**, and use them to locate the two drilled-out nuts on the two ground-out areas on the guide **A8**. Check that parts **A5** and **A8** are parallel; then weld the drilled-out nuts only to the guide (Fig. 6). Note that a small section of I-beam is useful for accurately clamping small parts ready for welding, as shown. Hold the I-beam in a vice, or clamp it to the bench. After welding, unscrew the bolts from the cross member **A5**.

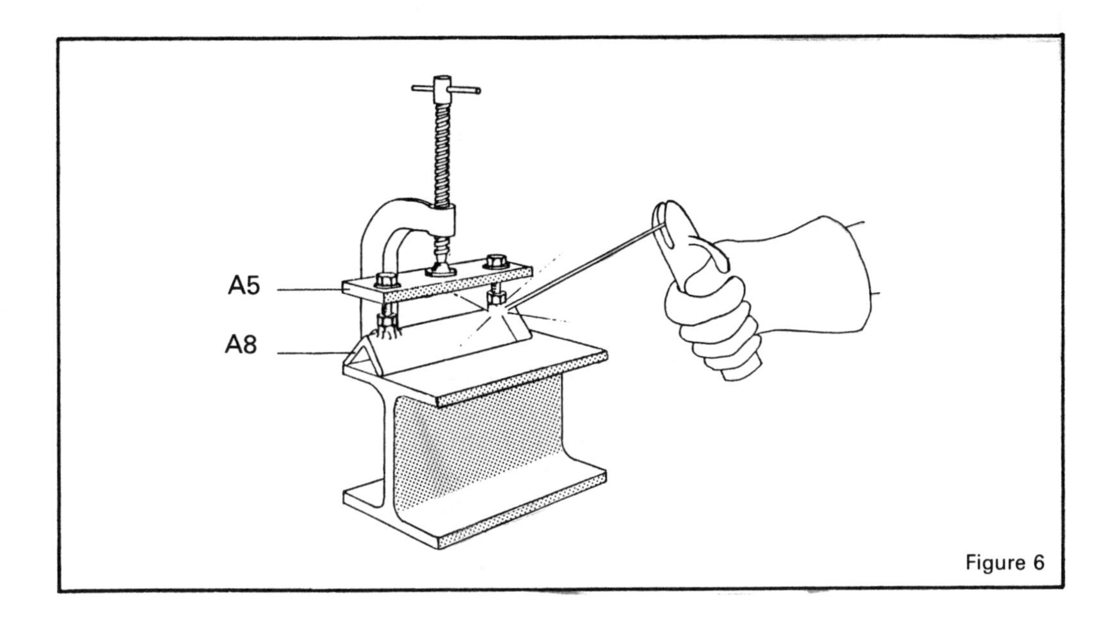

Figure 6

Lay out one frame side plate **A1**, with the pivot pin pointing down, and position the four cross members **A2, A3, A4** and **A5** on it as shown in Fig. 7. Check that **A4** and **A5** are at right angles to the top of the lower jaw. Tack weld on both sides of each cross member, check that they are still square and then weld fully. After the assembly has cooled, check again for distortion, and hammer straight if required.

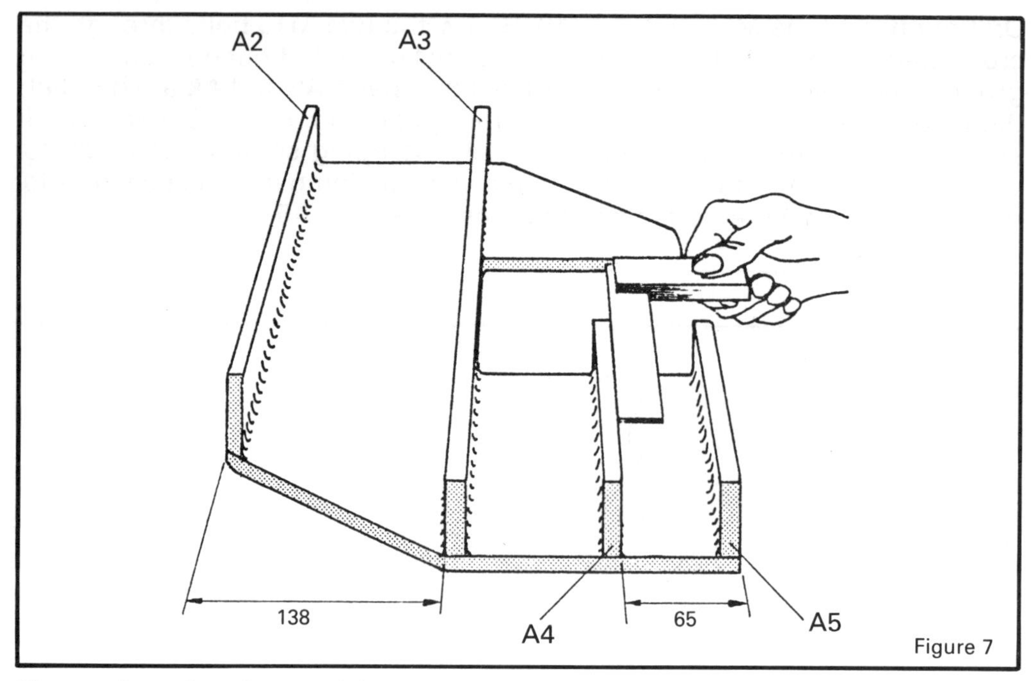

Figure 7

Clamp the other frame side plate **A1** in position, as shown in Fig. 8. Ensure that the pivot pin is pointing out from the frame. Check for straightness (Fig. 8), and weld the plate onto the four cross members. Again, check for distortion, and correct with a hammer if necessary. Clean off the weld spatter inside the box where the punch holder guide **A8** will be fitted.

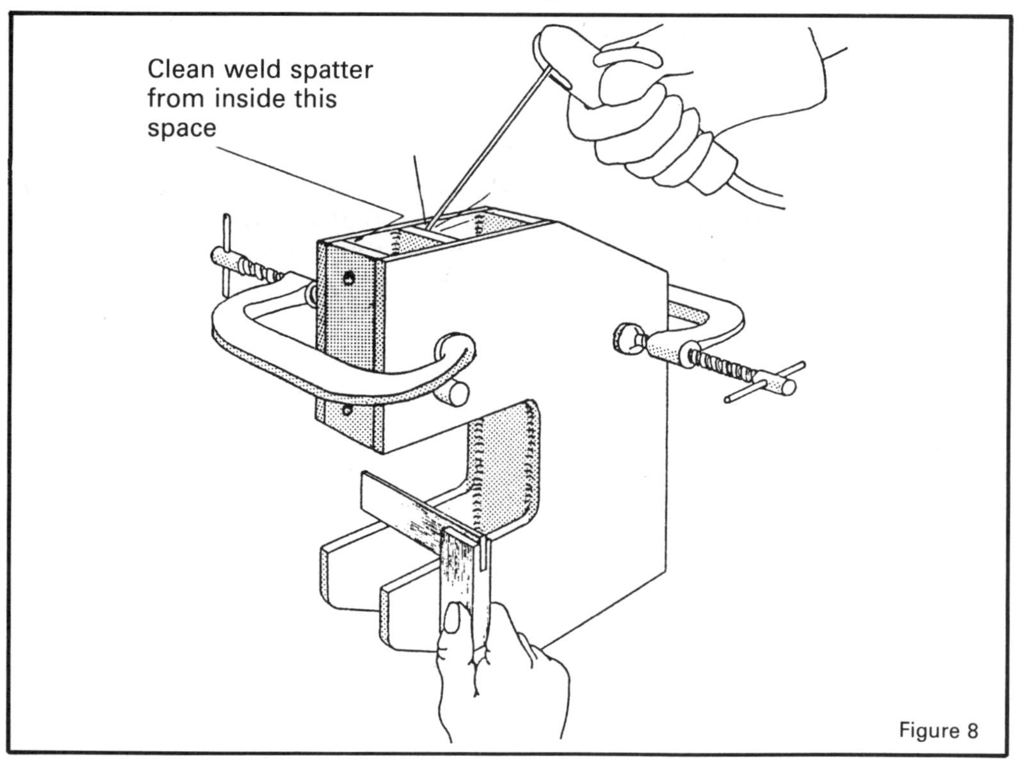

Figure 8

Tack weld the fillets **A6** and bench mounting flanges **A7** to the frame side plates. Check that they are all square, and then weld in short runs to complete the assembly (Fig. 9). Do not weld along the whole length, as this would cause the flanges to lift as the weld cools.

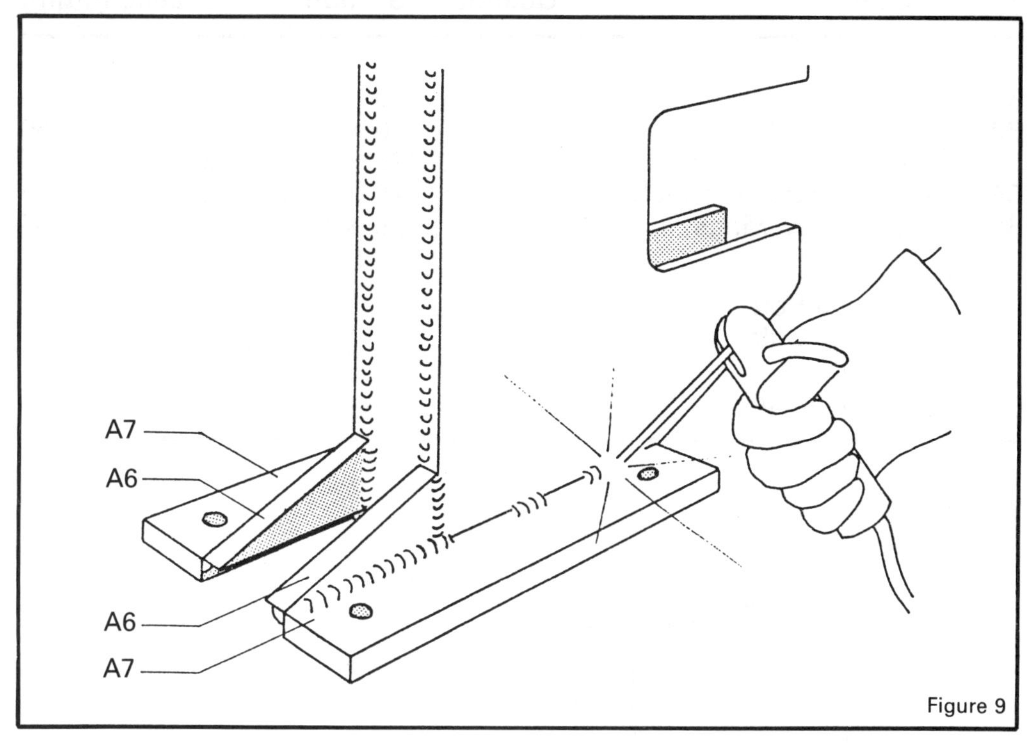

Figure 9

File the edges of the punch holder guide **A8** so that it is a close fit within the frame, between crossmembers **A4** and **A5** (Fig. 10). Assemble with the nuts and bolts, as shown.

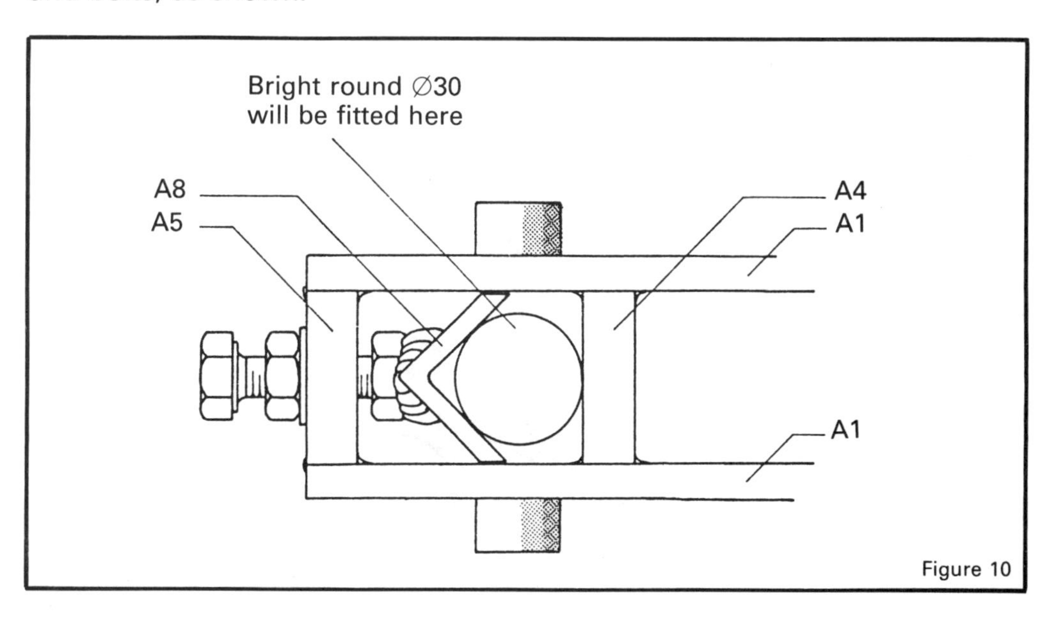

Figure 10

DIE MOUNTING — B
PARTS

Part	Name	Quantity	Section	Length(mm)
B1	Rear half plate	1	40 × 12 MS flat	60
B2	Front half plate	1	40 × 12 MS flat	60
B3	Rear handscrew shaft	1	25 × 5 MS flat	85
B4	Front handscrew shaft	1	25 × 5 MS flat	65
B5	Handscrew wings	2	25 × 5 MS flat	55
Fasteners				
	M8 × 30 Bolt	1		
	M8 × 25 Bolt	1		
	M8 Nut	1		
	M8 Flat washers	2		

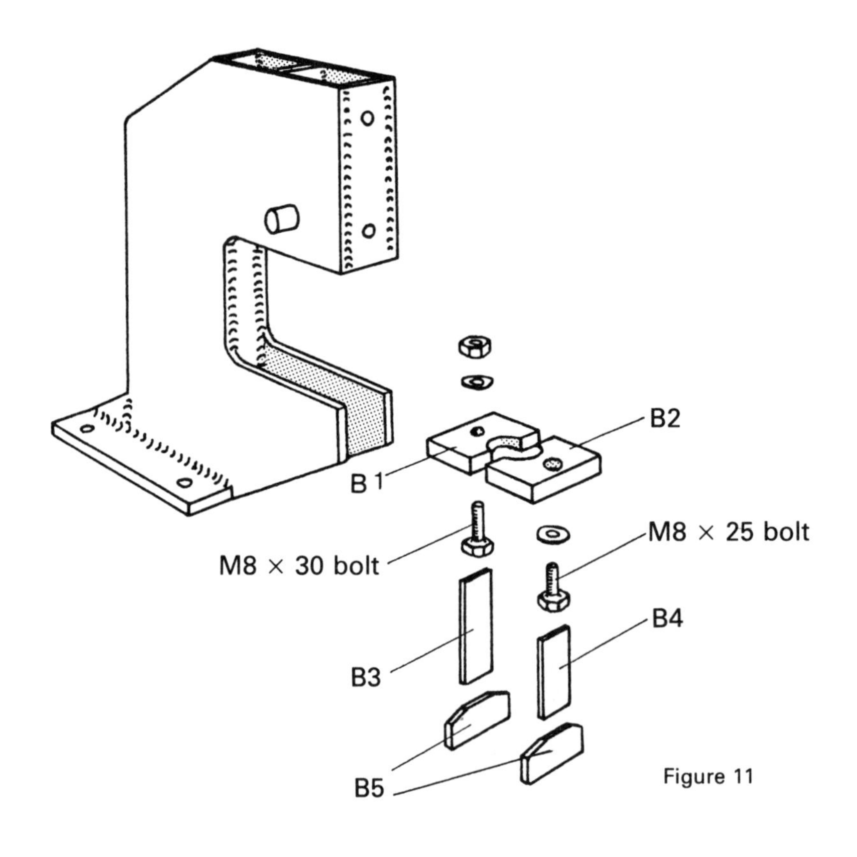

Figure 11

Cut the steel accurately for parts **B1** to **B5** to the dimensions given in the parts table. Carry out the additional cutting, drilling and tapping operations required for parts **B1**, **B2** and **B5**, as shown in Fig. 12. The ∅30 hole in parts **B1** and **B2** does not need to be cut with great precision; flame cutting would be sufficient.

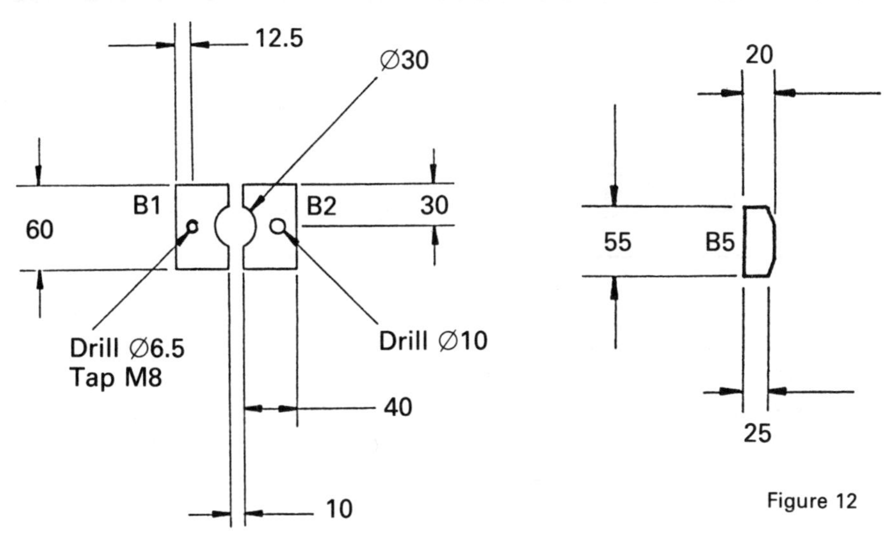

Figure 12

Lay parts **B1** and **B2** across the lower jaw of the frame; **B1**, with the M8 tapped hole, should be to the rear. Place a piece of ∅30 bright round bar through the frame (Fig. 13) and position **B1** and **B2** so that they fit closely round it. The bar can be held in place by the bolts through **A5**, pressing on the guide **A8**. Check that the M8 and ∅10 holes in **B1** and **B2** respectively are 65mm apart (centre to centre); check also that the front of **B2** is flush with the front of the lower jaw of the frame. Clamp **B1** and **B2**, check for squareness (Fig. 13), and then tack weld to the frame. Check again for squareness, remove the ∅30 bright round bar and weld **B1** and **B2** fully.

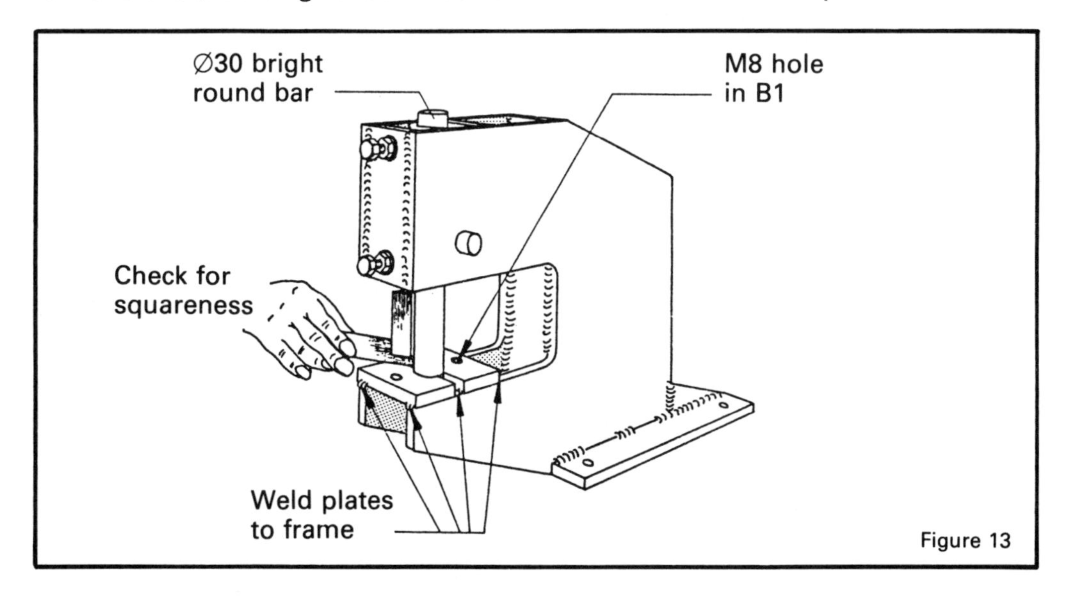

Figure 13

To make the rear handscrew, weld the M8 × 30 bolt, the shaft **B3** and one of the wings **B5** together (Fig. 14). Protect the thread of the bolt during welding with nuts or a piece of scrap metal. Make the front handscrew in a similar way using the M8 × 25 bolt, shaft **B4** and the other wing **B5**.

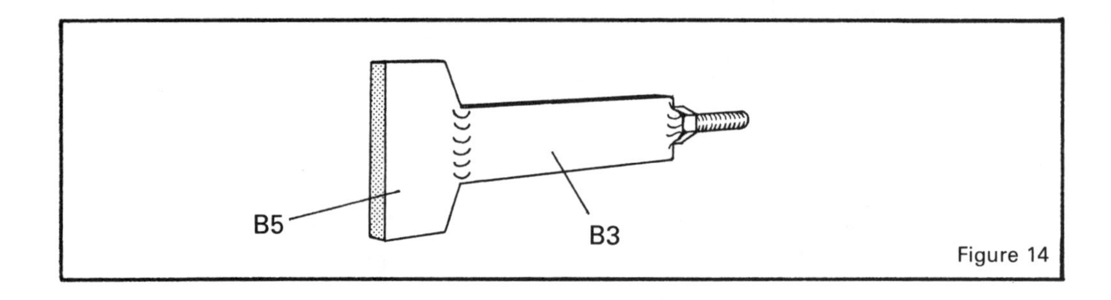

Figure 14

Screw the rear (longer) handscrew up through the M8 hole in **B1**. Place a washer and a nut over the protruding bolt, and tighten the nut until the top of the bolt is slightly below the top of the nut. Weld the nut, bolt and washer together (Fig. 15). Fit the front (shorter) handscrew with a washer, and store it for use later in the construction.

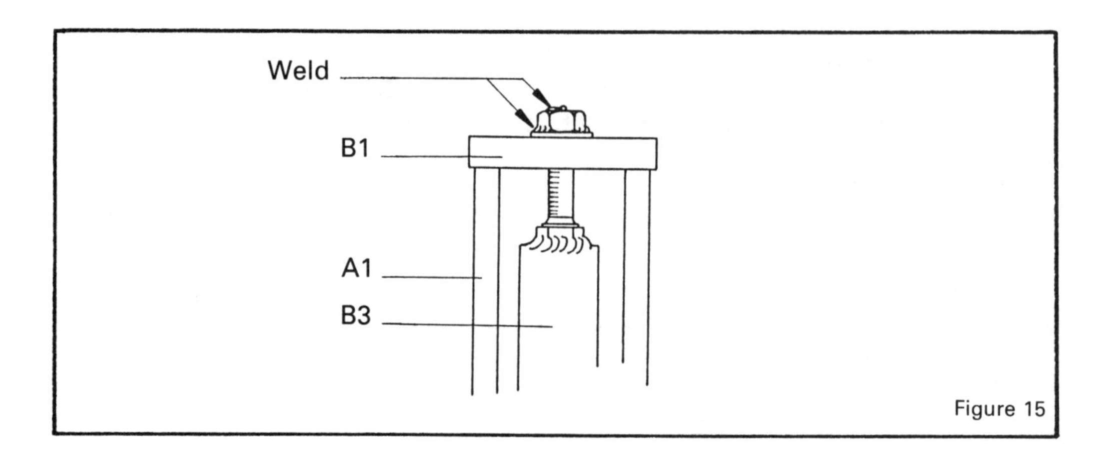

Figure 15

OPERATING MECHANISM — C
PARTS

Part	Name	Quantity	Section	Length(mm)
C1	Punch holder	1	∅30 bright round	260
C2	Punch holder reinforcing flats	2	40 × 12 MS flat	25
C3	Linkage plates	2	40 × 12 MS flat	200
C4	Counterweight bars	2	40 × 12 MS flat	300
C5	Fillets	2	40 × 12 MS flat (one piece cut diagonally into two)	110
C6	Handle mounting bars	2	40 × 12 MS flat	189
C7	Handle spacers	4	40 × 12 MS flat	15
C8	Handle	1	25mm nominal bore medium duty tube	600
C9	Punch holder operating pin	1	∅20 bright round	125
C10	Linkage operating pins	2	∅20 bright round	32
C11	Handscrew wing	1	25 × 5 MS flat	40
Fasteners				
	M10 × 25 Bolts	4		
	M10 Flat washers	4		
	M8 × 20 Bolt	1		
	Split pins	4	3 × 30	

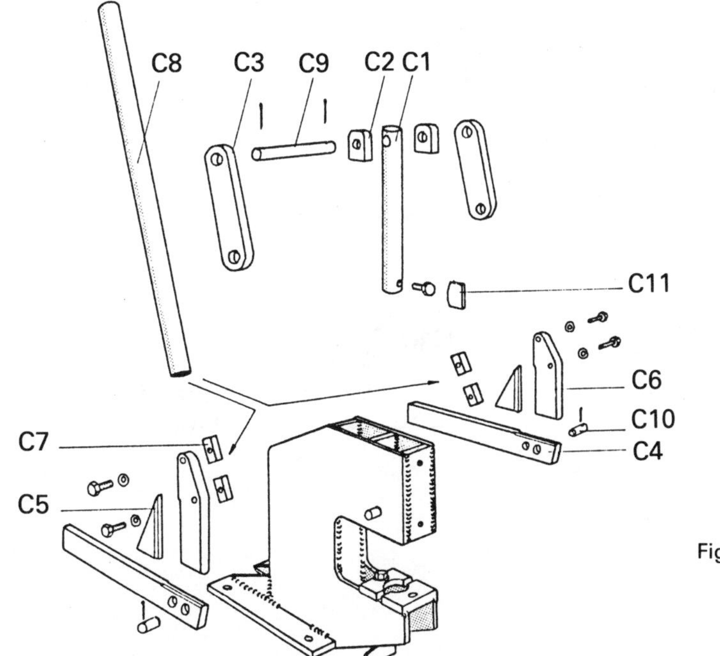

Figure 16

Cut the steel accurately for parts **C1** to **C11** to the dimensions given in the parts table. Carry out the additional cutting, drilling and tapping operations shown in Fig. 17.

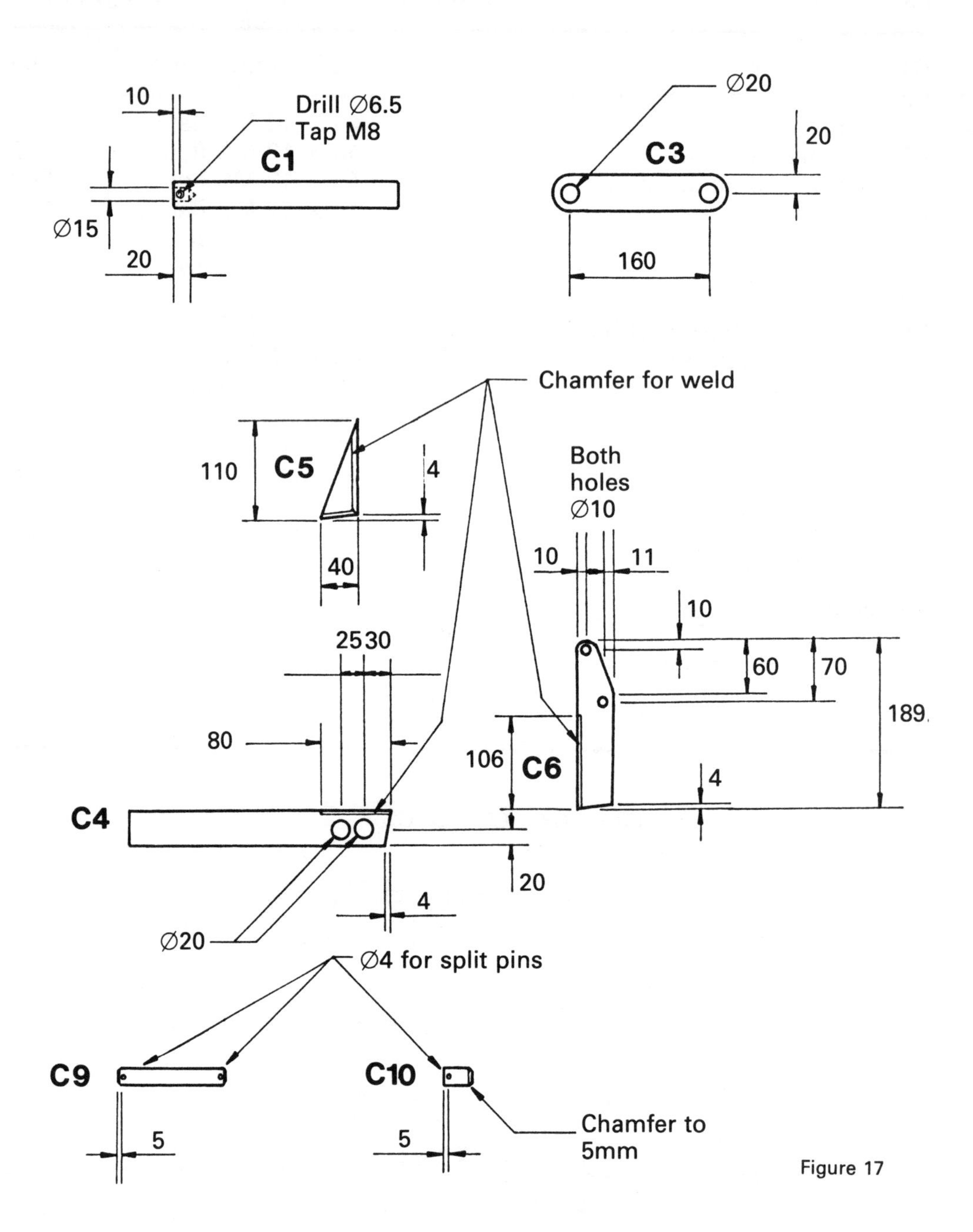

Figure 17

Clamp the punch holder **C1** to a flat surface with the tapped hole vertical. Clamp the reinforcing flats **C2** to the sides at the opposite end from the tapped hole, as shown in Fig. 18. Check that both flats **C2** are parallel, and then weld them to the punch holder.

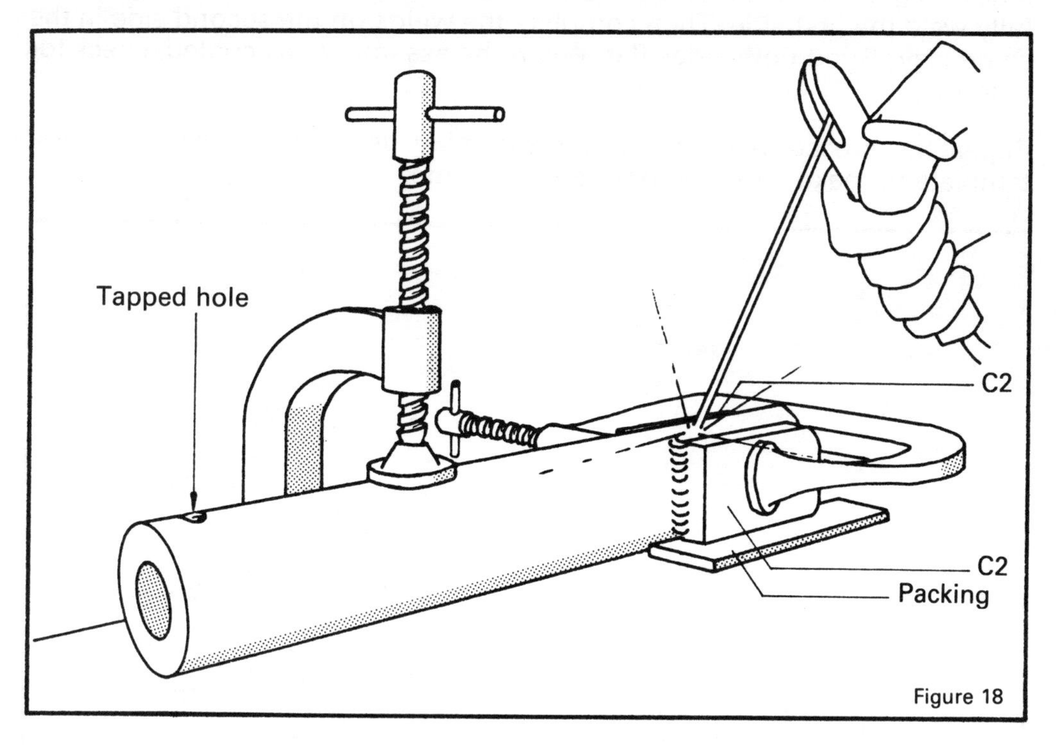

Figure 18

Drill a ∅4mm hole through parts **C1** and **C2**, then enlarge to ∅20mm, as shown in Fig. 19. The hole must be at right angles to the axis of the punch holder.

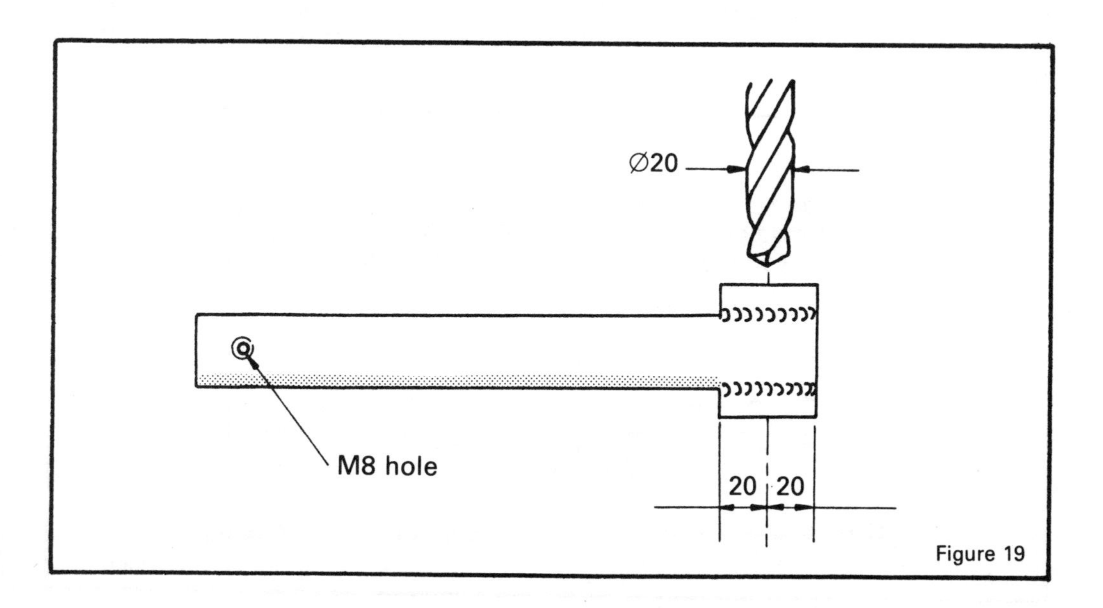

Figure 19

Take parts **C4**, **C5** and **C6** for one counterweight lever assembly, arrange them as shown in Fig. 20 and clamp them to a flat piece of metal. Tack weld them together. Grind the tack welds flat, then turn the assembly over, clamp it down again and tack weld the other side strongly. Turn it over again and fully weld the first side. Then complete the welds on the second side in the same way. Grind both sides flat. When the assembly has cooled, check for flatness.

Repeat this procedure for the second assembly, using the first assembly as a template to make sure that both are the same.

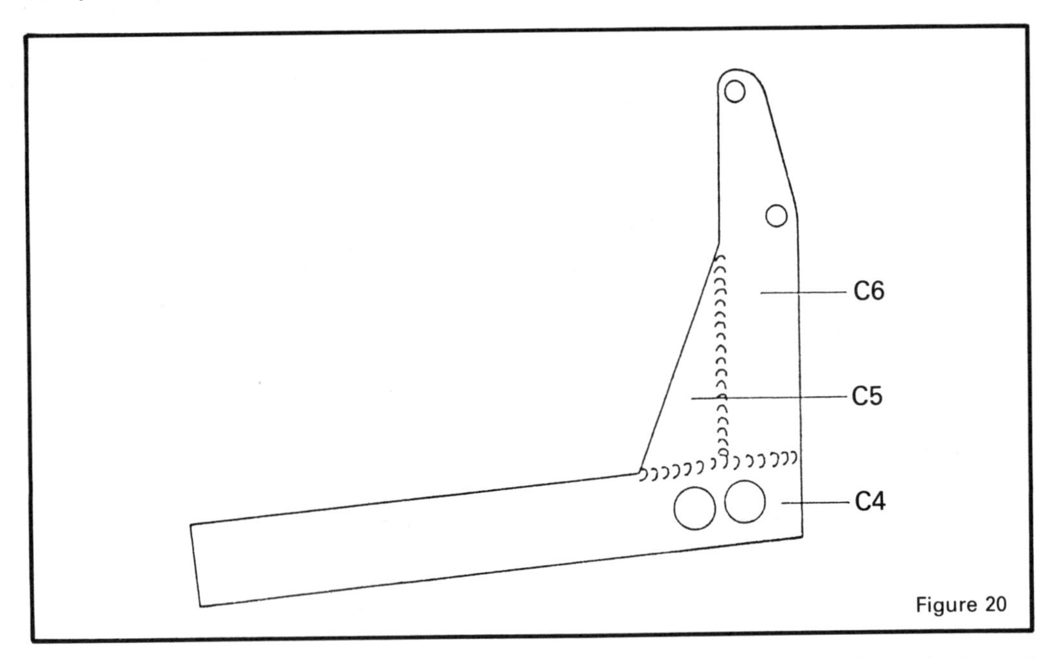

Figure 20

The two linkage operating pins **C10** are now welded into the front holes of the assemblies (Fig. 21). Take care to have the pins protruding on opposite sides, as shown.

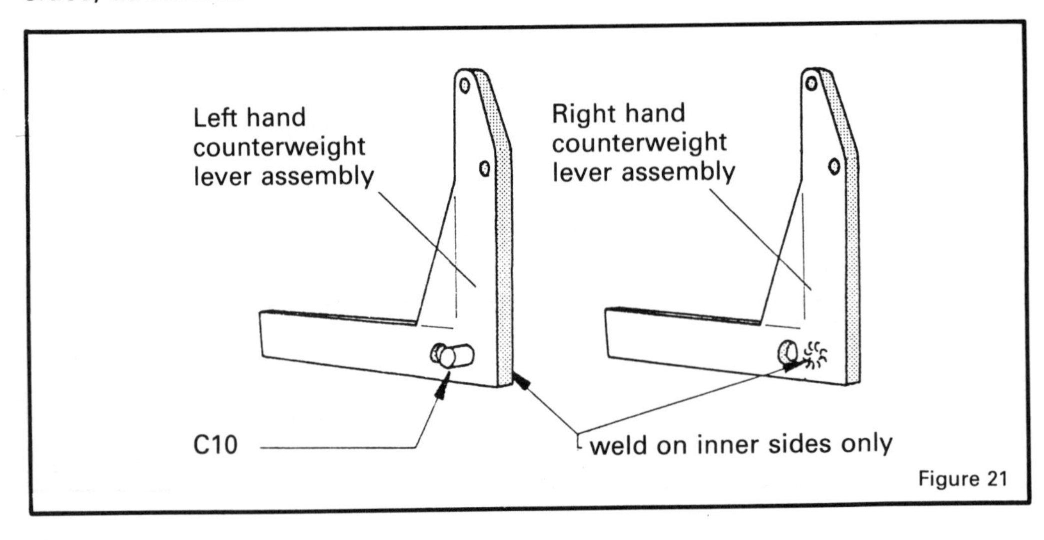

Figure 21

The handle **C8** is made from a piece of tube; one end may first be filled in and rounded to provide a smooth surface for the operator's hand.

Grease the two linkage pivot pins **A9**, and fit the two counterweight lever assemblies onto them, as shown in Fig. 22. Place thin sheet metal (1mm approx.) between the assemblies and the frame, to provide clearance for paint etc., and clamp in the upright position. Drill each of the handle spacers **C7** centrally with a ∅8.5 hole, and tap to M10. Bolt them to the lever assemblies with the four M10 × 25 bolts and washers. Clamp the handle **C8** in place, and weld it to the spacers. When the welds have cooled, remove the clamps and thin sheet metal; check that the assembly moves freely on the pivot pins **A9**.

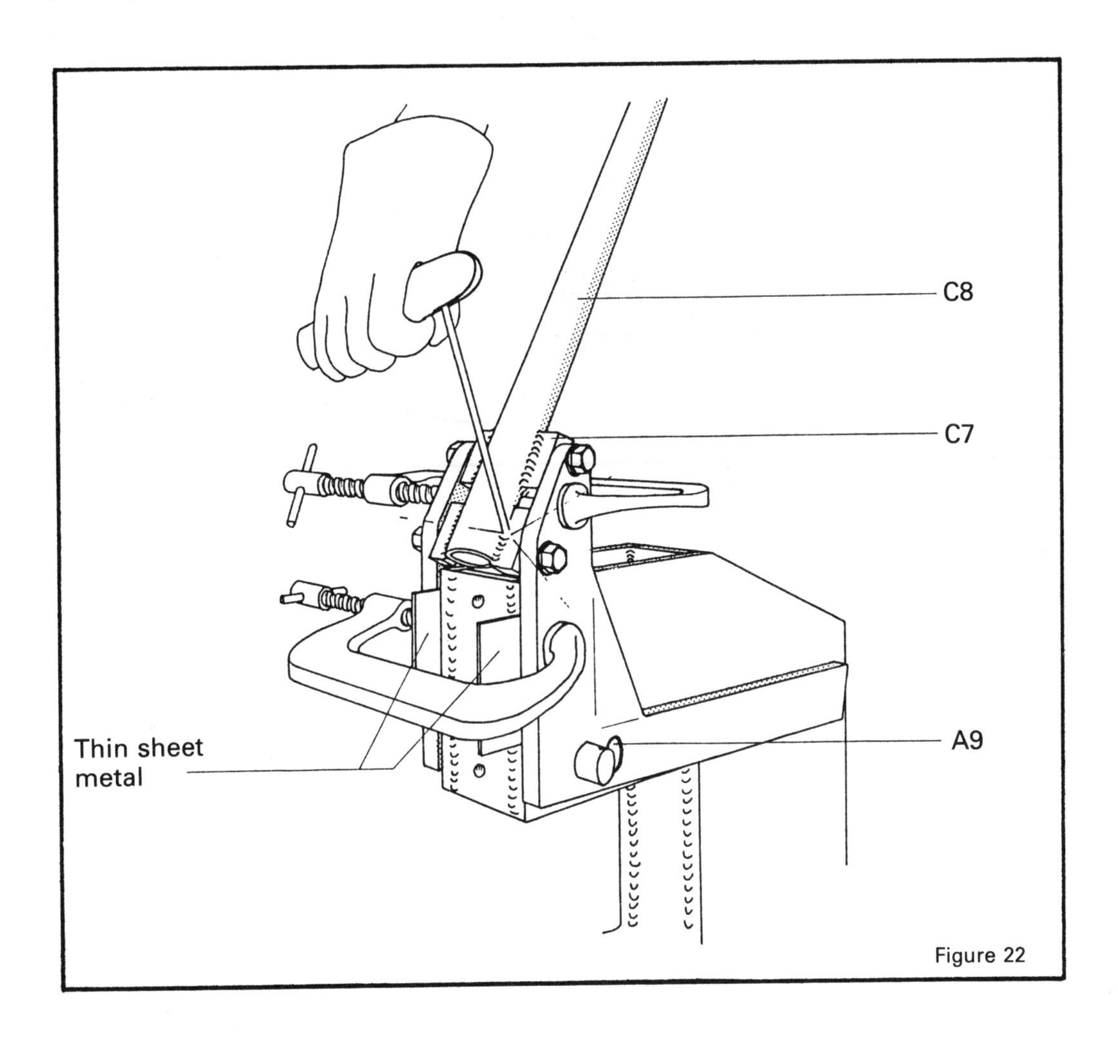

Figure 22

Grease and fit the punch holder **C1** into the frame, with the M8 hole at the bottom, and facing forwards. The holder is located by the guide **A8**, which is adjusted to give a good sliding fit by the two M10 bolts in **A5**. Grease and fit the punch holder operating pin **C9**, through the ∅20 hole drilled in **C1** and flats **C2**. Fit the two linkage plates **C3** between **C9** and the punch holder operating pins **C10**. Split pins are used to hold these linkages in place. Check that the assembly operates smoothly.

Weld the handscrew wing **C11** to the M8 × 20 bolt, as shown in Fig. 23. Fit the handscrew into the M8 hole in the front of the punch holder **C1**.

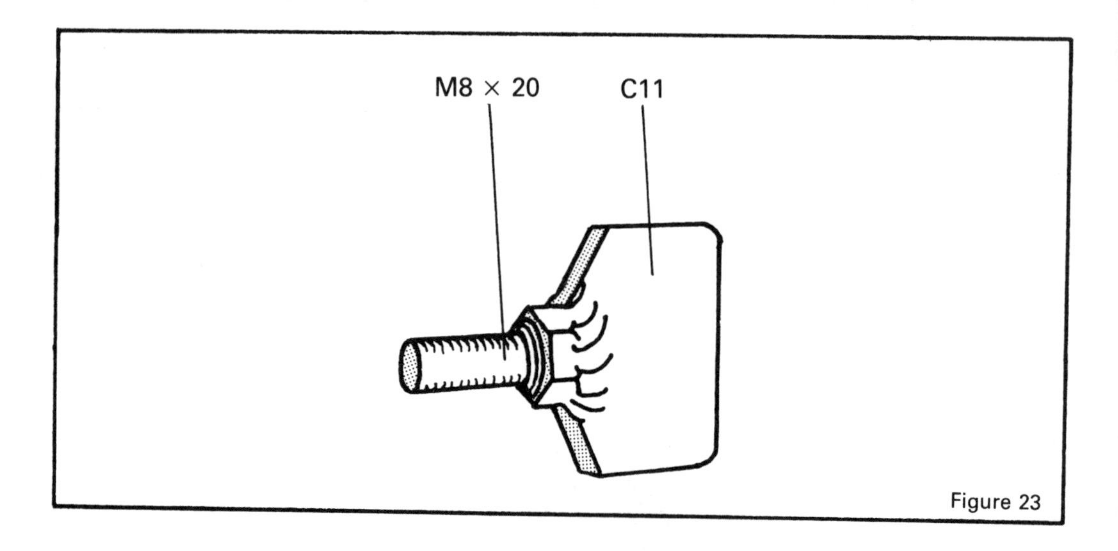

Figure 23

STRIPPERS — D
PARTS

Part	Name	Quantity	Section	Length(mm)
D1	Stripper arms	2	25 × 5 MS flat	80
D2	Counterweights	2	25 × 5 MS flat	30
Fasteners				
	M8 × 30 Bolts	2		
	M8 Nuts	4		

Cut the steel accurately to the dimensions shown in the Parts table for parts **D1** and **D2**. Cut the heads off the two M8 bolts and cut both in half. Drill two of the M8 nuts out to ∅8. Bend the stripper arms to the dimensions shown in Fig. 24.

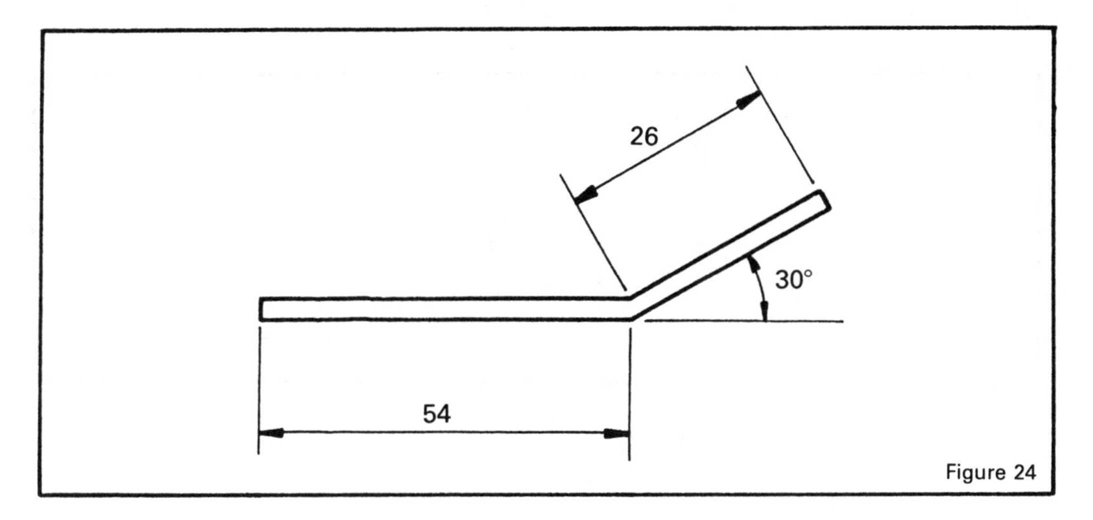

Figure 24

Clamp the halves of the M8 bolts with one of the arms **D1**, as shown in Fig. 25. The threads must be protected with the M8 nuts, and packing pieces must be used to ensure that all the parts line up correctly. Check that the two bolt halves are in line, and square with **D1**, and weld together. Repeat for the other arm **D1**.

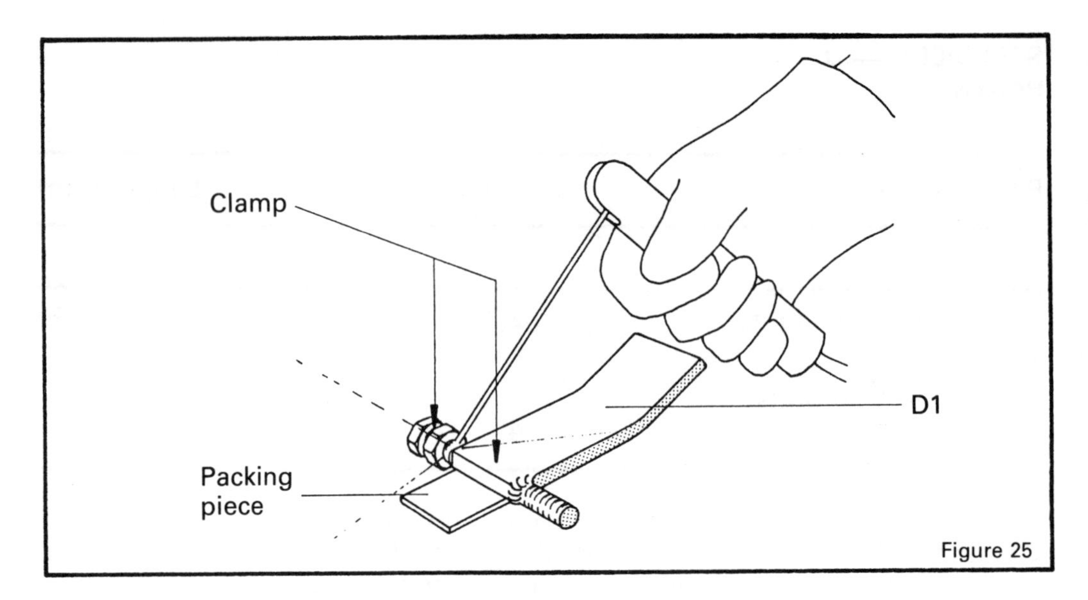

Figure 25

Weld the counterweights **D2** to the arms **D1**, at right angles to the upper (longer) part. Grind the surface of each arm as indicated in Fig. 26.

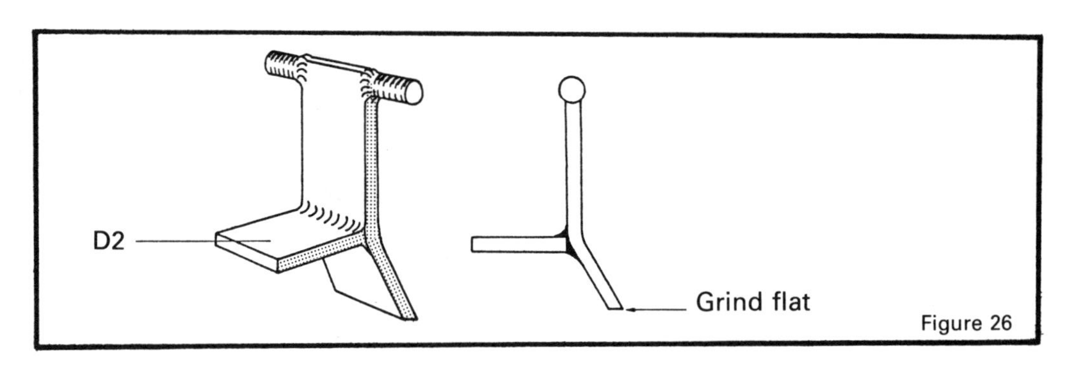

Figure 26

Thread one of the M8 nuts onto the rear bolt half of each stripper arm assembly, so that the nut is in the centre of the threaded portion. Place one of the drilled-out nuts on each of the front bolt halves of each arm assembly, again positioning it centrally on the threaded portion. Clamp both stripper arm assemblies to the punch holder, with packing pieces 2mm thick on either side of the punch holder (Fig. 27). Weld the nuts to the frame, taking care that the welds do not stand out far enough to interfere with the operating mechanism. Grind the welds if necessary. Ensure also that the stripper arms swing freely when vertical, so that different sizes of punch can be accommodated. However, friction when the arms are up in the horizontal position is helpful, as punches can then be changed more easily.

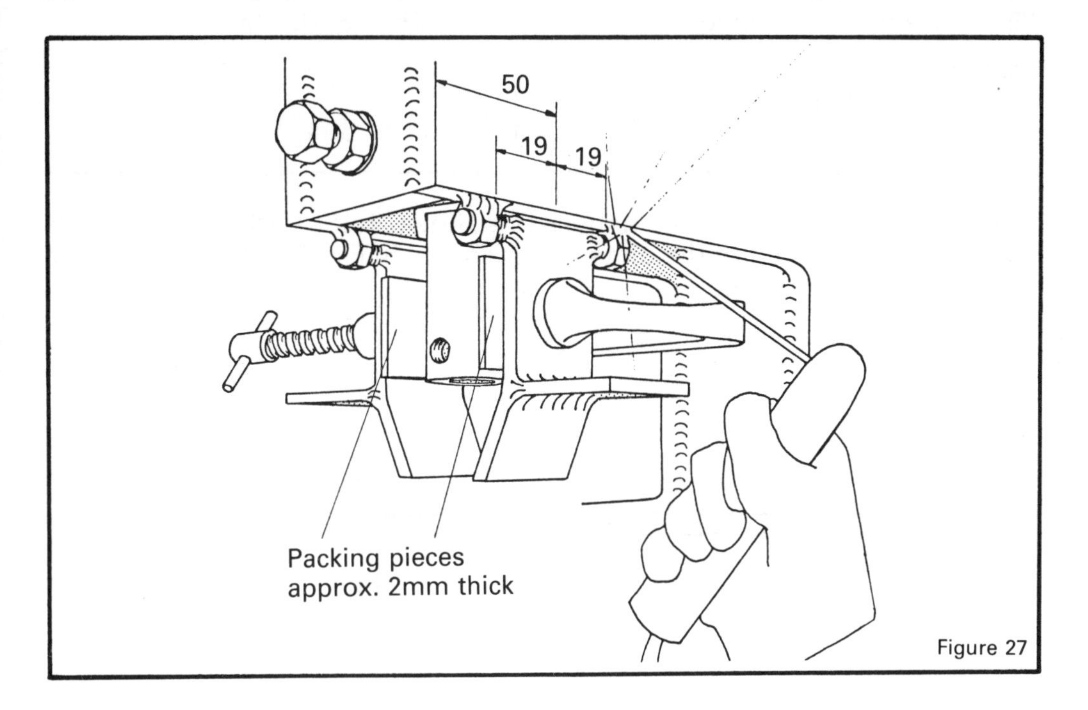

Figure 27

DEPTH GAUGE — E
PARTS

Part	Name	Quantity	Section	Length(mm)
E1	Arm	1	25 × 5 MS flat	185
E2	Face piece	1	25 × 5 MS flat	28
E3	Clamp spacers	2	25 × 5 MS flat	75
E4	Clamp plate	1	25 × 5 MS flat	65
E5	Clamp screw wing	1	25 × 5 MS flat	65
Fasteners				
	M8 × 20 Bolt	1		
	M8 Nut	1		

Cut the steel accurately to the dimensions shown in the Parts table for parts **E1** to **E5**. Bend arm **E1** to the dimensions shown in Fig. 28, making sure that the bend is a right angle.

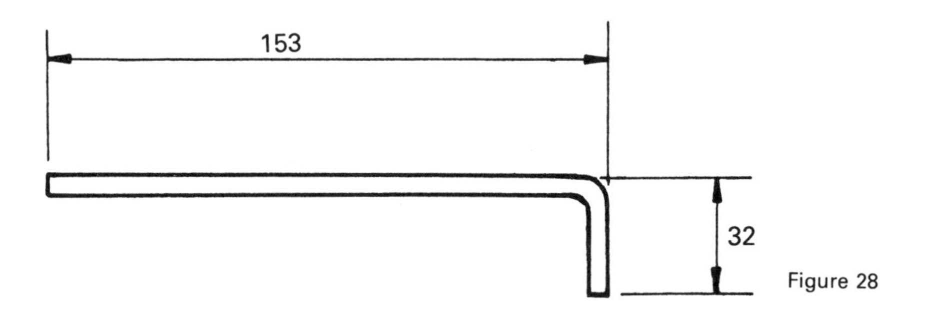

Figure 28

Weld face piece **E2** to arm **E1**, making sure that the front face is flat, and that **E2** is at right angles to the longer part of **E1** (Fig. 29). Grind the weld if necessary.

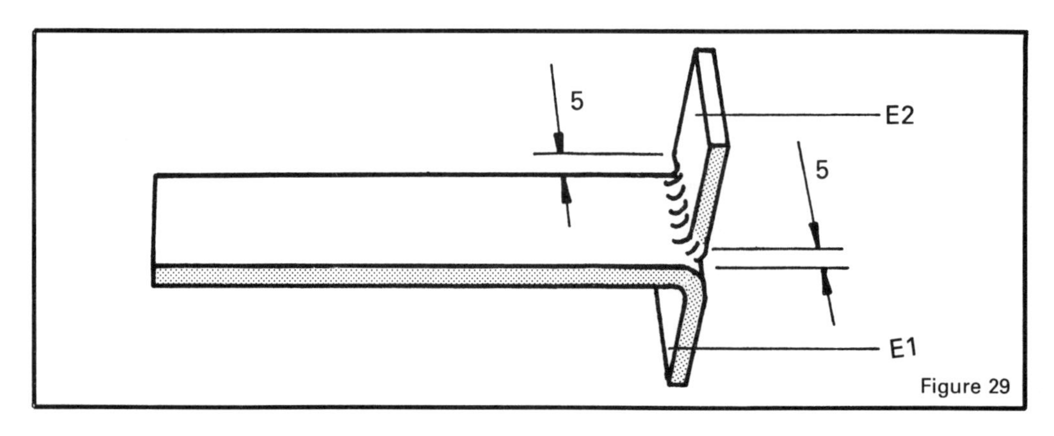

Figure 29

Drill a ∅8.5 hole in the centre of clamp plate **E4**. Bend the clamp plate **E4**, and weld the M8 nut over the ∅8.5 hole, making sure that the M8 × 20 bolt will pass squarely through both the nut and the hole (Fig. 30). Round all the corners on clamp screw wing **E5**. Weld the wing **E5** to the M8 × 20 bolt, to form a handscrew, as in previous sections.

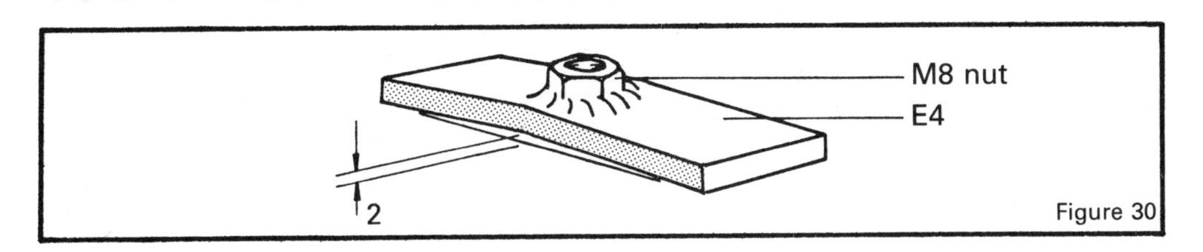

Figure 30

Lay the frame on its right side (as viewed from the front), with the left hand side uppermost. Position the components and weld as shown in Fig. 31. After welding, check that the gauge assembly (**E1**, **E2**) slides freely between parts **E3** and **E4**; file or emery the edges of **E1** as required. Screw the handscrew in through the M8 nut to clamp the gauge in place.

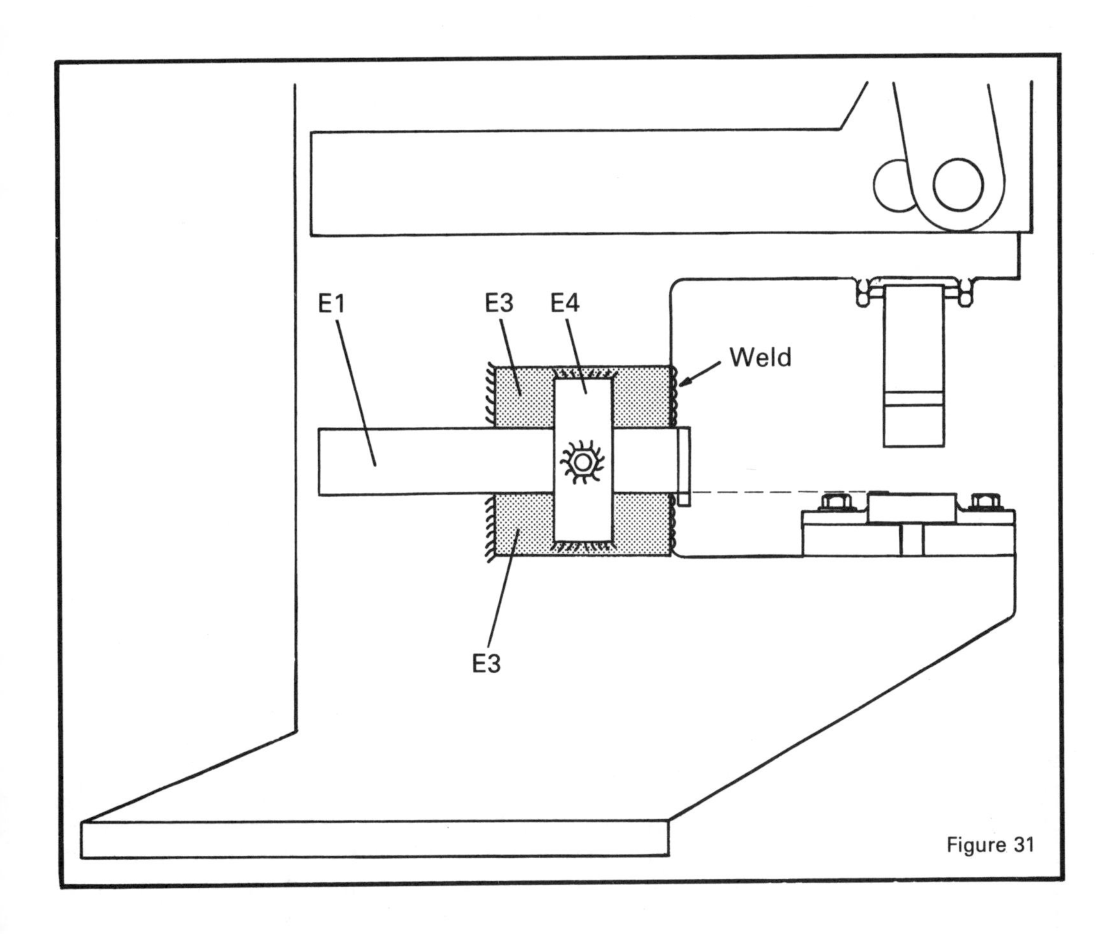

Figure 31

PUNCH AND DIE — F
PARTS

Part	Name	Quantity	Section	Length(mm)
F1	Rear die mounting plate	1	25 × 5 MS flat	60
F2	Front die mounting plate	1	25 × 5 MS flat	60
F3	Die	1	12 × 60 (nominal) vehicle leaf spring	40
F4	Punch	1	∅15 min. high carbon steel (see text)	40
Fasteners				
	M8 Nut	1		

Cut the steel accurately to the dimensions shown in the Parts table for parts **F1**, **F2** and **F3**. Carry out the additional drilling and cutting operations required for **F1** and **F2**, as shown in Fig. 32.

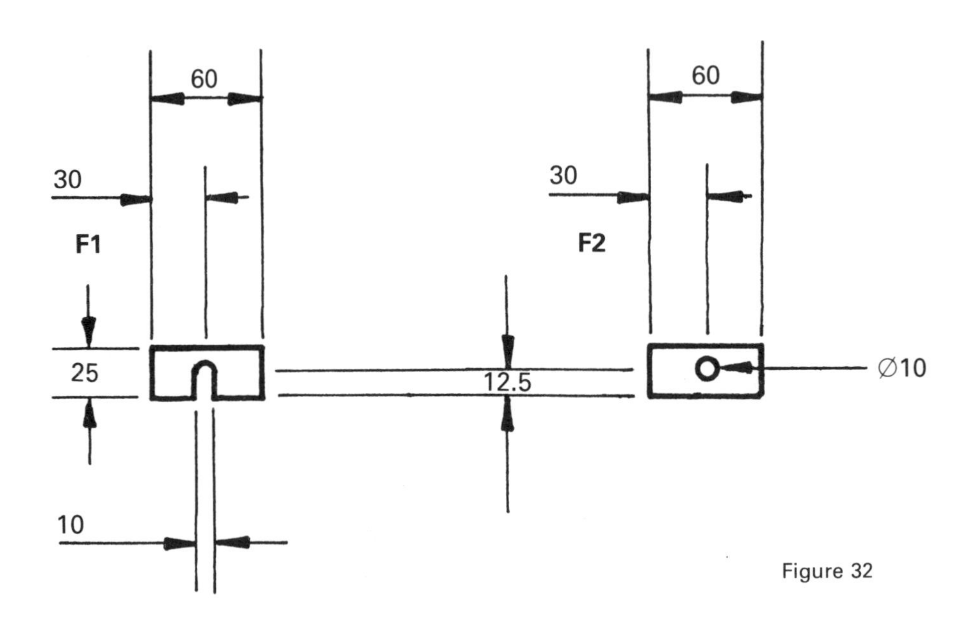

Figure 32

Soften the piece of leaf spring to be used for the die **F3**. This can be achieved by heating it until it is dull red, and then leaving it to cool slowly, preferably burying it in sand while it is cooling. Drill a hole through the centre of the die, of the same diameter as that required for the punch. Now drill this hole out to a diameter about 4mm larger than the existing one, through *half the thickness* of the die only (Fig. 33). This will allow the blanks to drop through the die, once they have been punched.

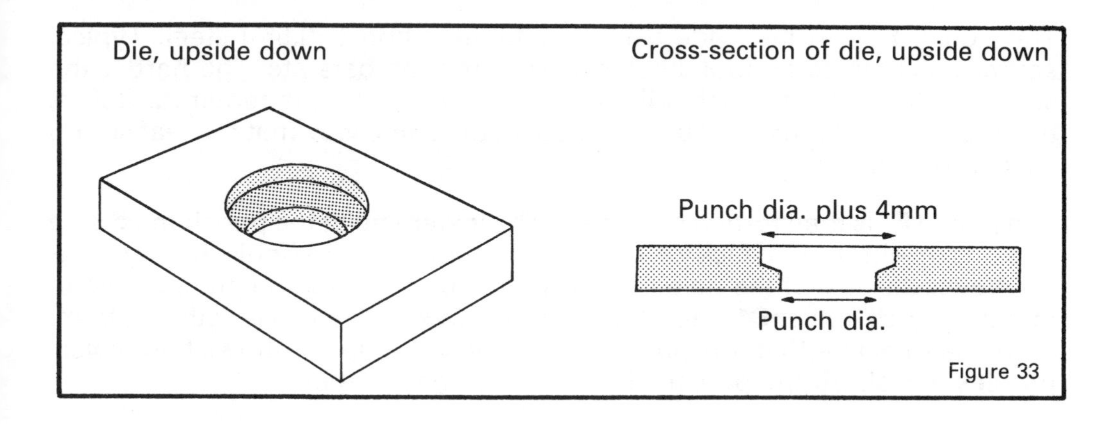

Figure 33

If a drill bit of the right size is not available, drill out the die to the nearest smaller size for which a bit is available. Turn down about 5mm on the end of a piece of scrap bar to fit this hole (Fig. 34), and tack weld the bar to the die. Clamp the bar in the chuck of a lathe, and bore out the die to size. Afterwards, the tack weld can be ground or sawn through, to remove the die. Grind flat any weld remaining.

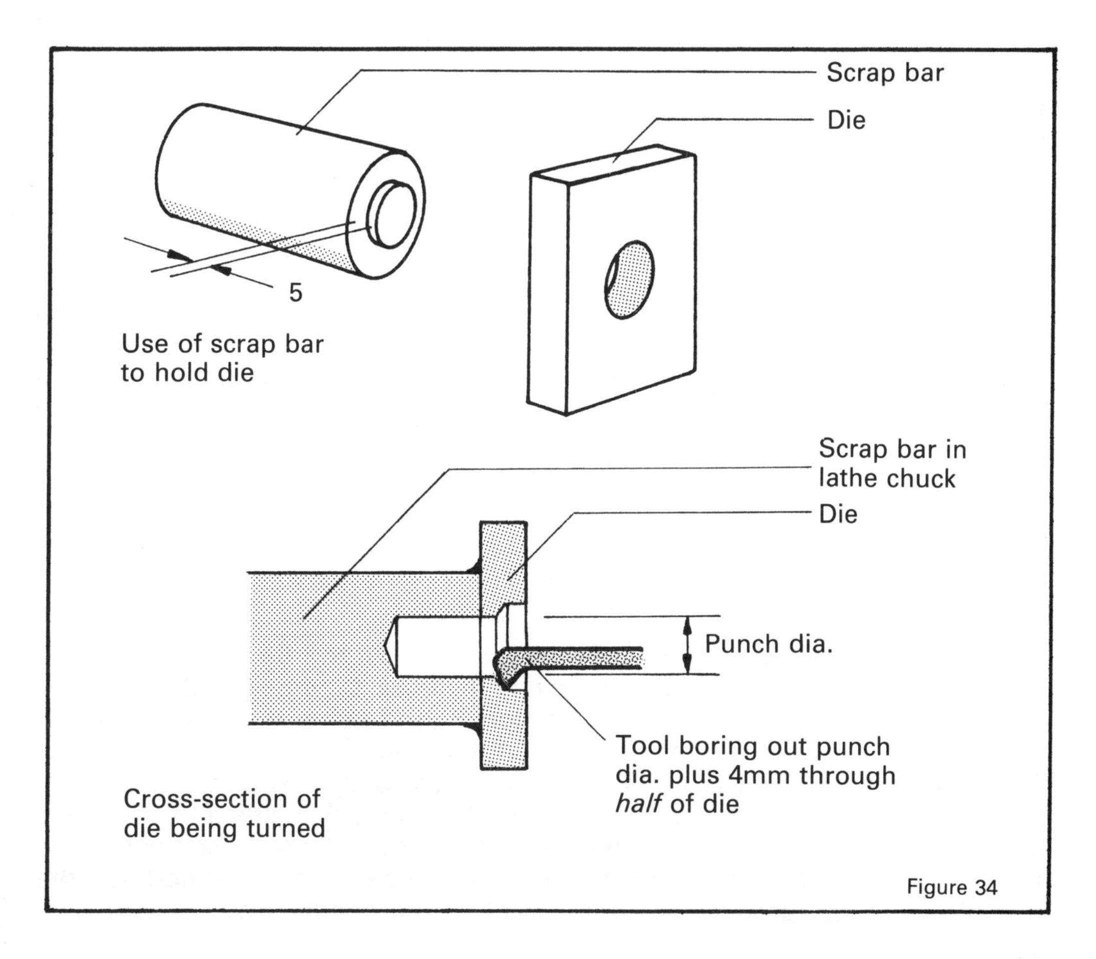

Figure 34

The punch **F4** can be made from any high-carbon (>0.5%) steel. Typical sources include scrap tool steel, drill bits, torsion bars etc. The harder the steel, the longer the punch will last before sharpening is required. Before turning, the steel must first be softened, in the same way that the leaf spring was softened.

The punch **F4** is then turned on a lathe. The lower end of the punch **must** be a close fit in the upper face of the die. The lower end should be concave. Whatever the size of the hole required, the upper end of the punch must fit into the punch holder **C1**; therefore, it must have ∅15 over a length of 20mm. Also, the total length of the punch must be 40mm. Fig. 35 shows two typical profiles. Finish off the punch with a file or emery paper.

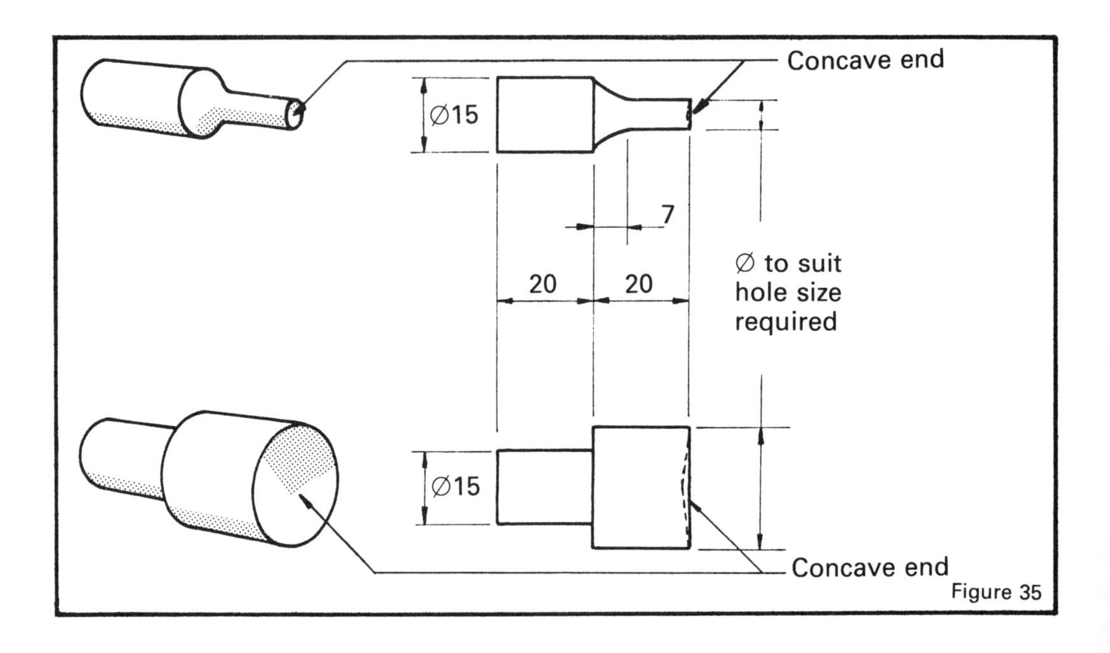

Figure 35

Place **F1**, **F2** and **F3** on the die mounting plates **B1** and **B2** in the frame. **F1** should be at the rear, with the slot open to the back. Clamp **F1** and **F2** in place with the handscrews made up in Section B, and using the M8 nut on **F2**. Insert the punch in the punch holder **C1**, clamping it with the handscrew. Position the die **F3** on **F1** and **F2** by bringing the punch **F4** down through it. Clamp the die and weld it to **F1** and **F2** with good tack welds (Fig. 36). Take care not to weld **F1** to the rear handscrew. Weld the M8 nut to **F2**, taking care not to weld it to the front handscrew. After the welds have cooled, the die assembly can be removed by loosening the two handscrews.

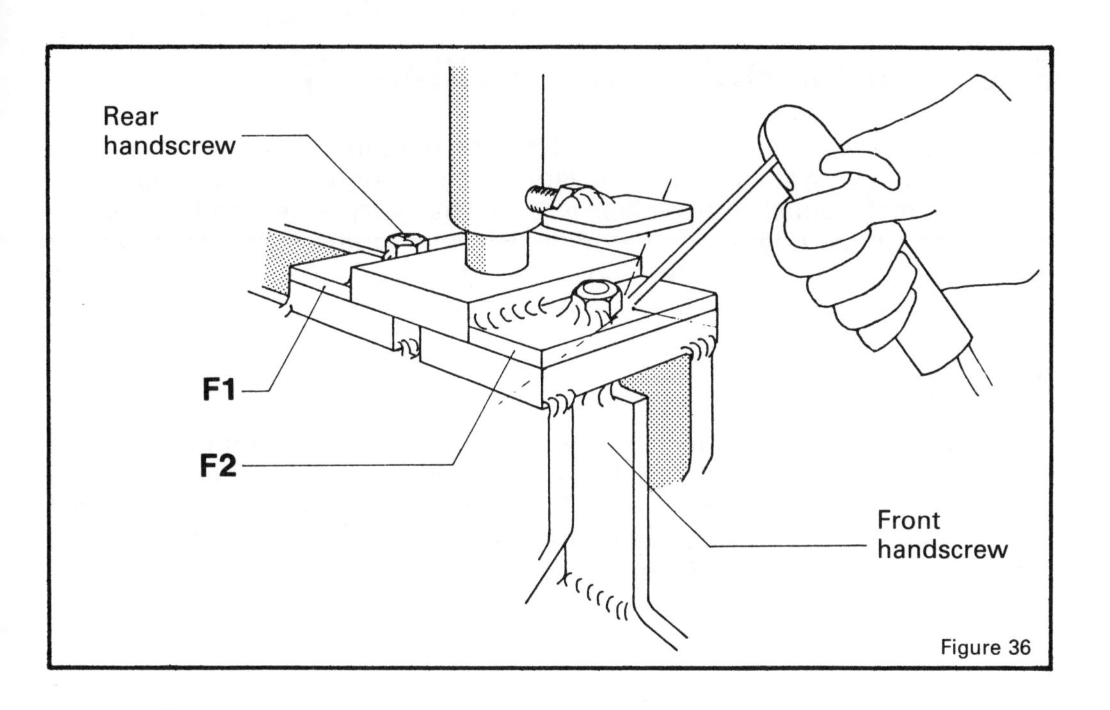

Figure 36

The punch and die must now be re-hardened, by heating to a bright red, and quenching rapidly in a bucket of waste oil. Do this in an open area, as the oil may burst into flames, and give off thick black smoke. Do not use water for quenching, as the steel may crack.

After hardening, grind a groove in the punch to locate the handscrew; also, grind a groove in the cutting end, to make the starting of the hole easier (Fig. 37).

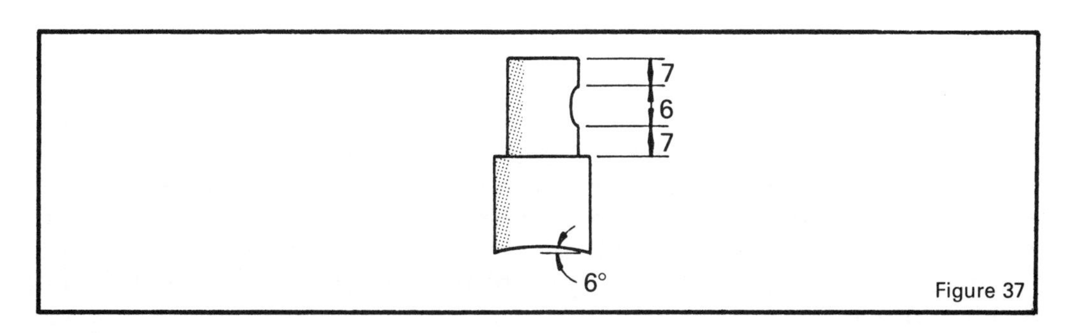

Figure 37

Rectangular Holepunch ('Nibbler')

Punches can also be made which have a rectangular shape. By making a series of holes with such a punch, a very large rectangular hole can be made. If one side of the punch is also given a curve, then very large circular holes, or holes with curved edges, can be made. However, these punches, and their dies, cannot be made by turning on a lathe, and other techniques are required. Two such techniques are described below.

Welded punch technique

Make the shank for the punch by softening and turning a 25mm length of high carbon steel down to ∅15. Make the punch itself by cutting a piece of leaf spring to the approximate size of the hole required; it needs to be about 15mm deep. Fig. 38 shows the dimensions for a punch to cut a hole 20 × 12, with one side having a ∅60 curved edge.

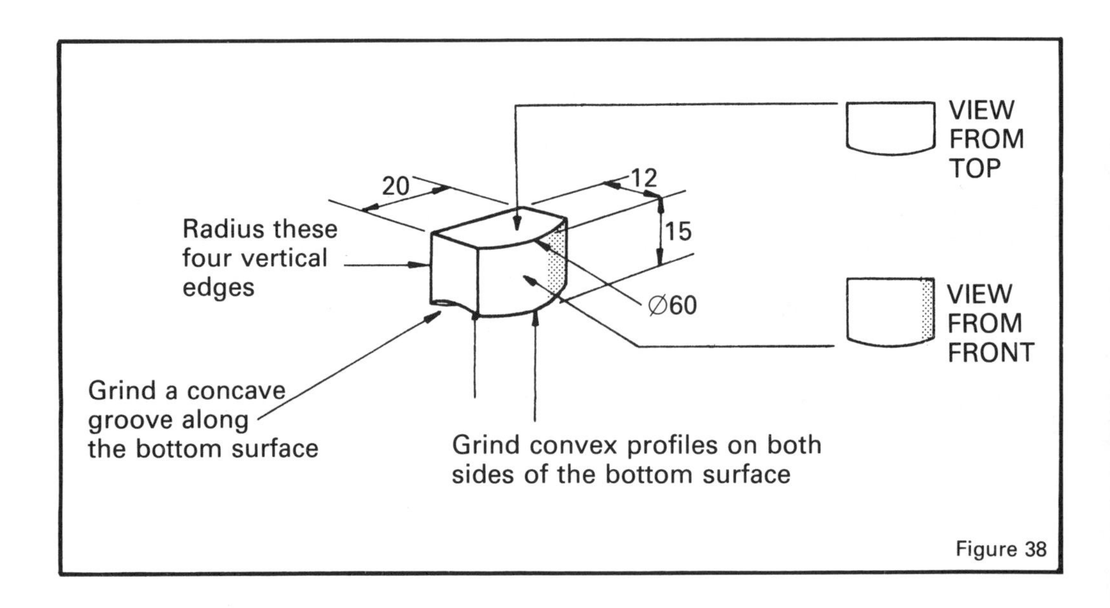

Figure 38

These dimensions can be formed by grinding. The inside radius of a pipe or bearing shell can be used as a guide when grinding the curved edge. The four vertical corners should be rounded off; otherwise, they will wear very rapidly. The bottom of the punch is hollow ground lengthwise and the cutting edges radiused lengthwise so that the two middle points will cut first. It is most important that the punch be symmetrical, and great care should be taken to ensure this; otherwise, the finished punch and die set will not function properly.

Weld the shank to the punch, ensuring that the centre line of the shank is in line with the centre of the punch (Fig. 39). After welding, heat the assembly to a dull red heat and allow it to cool slowly, in order to relieve any stresses. Check that the punch is symmetrical and parallel with the shank. File flats in both sides of the shank to accommodate the end of the handscrew. Harden the punch by heating to a bright red and quenching in oil, as described previously.

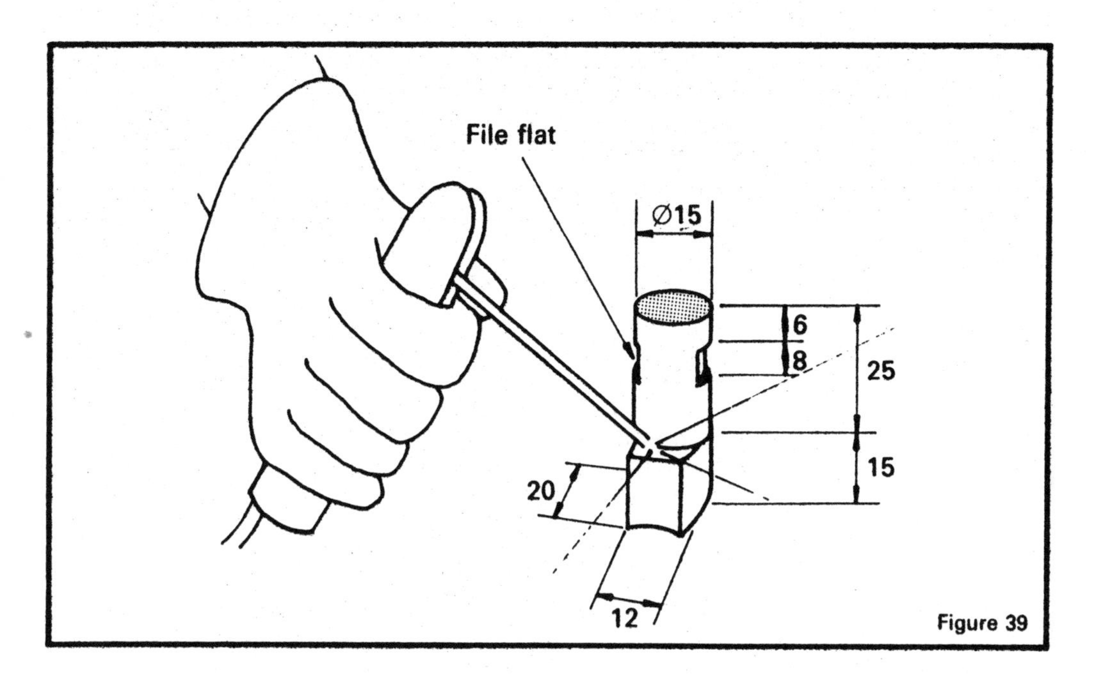

Figure 39

Forged punch technique

Soften and turn a 40mm length of high carbon steel down to have a Ø15 shank at one end, and a diameter larger than the largest dimension of the hole required at the other end (Fig. 40). For the 20 × 12 hole, as described above, Ø22 would be suitable. Then forge and grind the punch to achieve a similar form to that shown in Figs. 38 and 39. However, the meeting of the shank and the punch should be radiused, rather than made into a sharp step. Harden the punch.

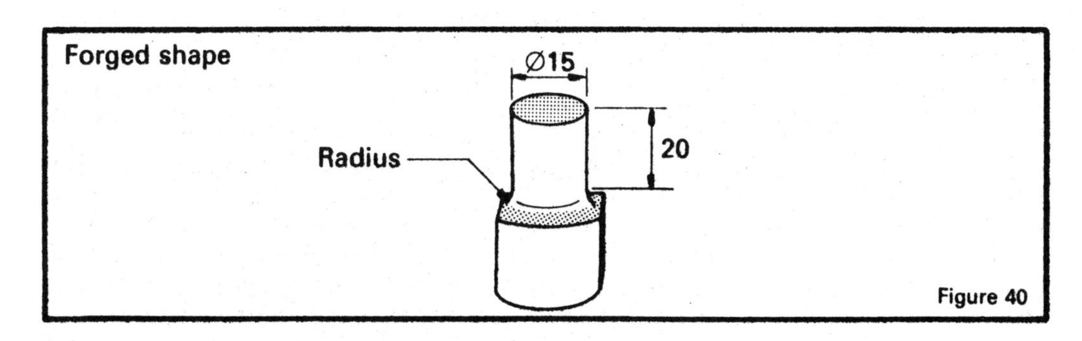

Figure 40

DIE MANUFACTURE

The die and its mounting plates are first cut to the same dimensions as those given in Section F. However, both **F1** and **F2** should be made with slots, so that the die can be mounted either way round. Soften the die **F3**, and drill a Ø3 hole in the centre. Then drill *halfway* through the centre of the die with a drill 4mm larger in diameter than the diagonal measurement of the hole to be punched. For the 20 × 12 hole, this would be approximately 31mm.

Heat the die to a bright red, and lay it with the large hole uppermost over a metal bolster or anvil with a hole just larger than the punch. Hammer the hardened punch through the die to form the shaped hole (Fig. 41).

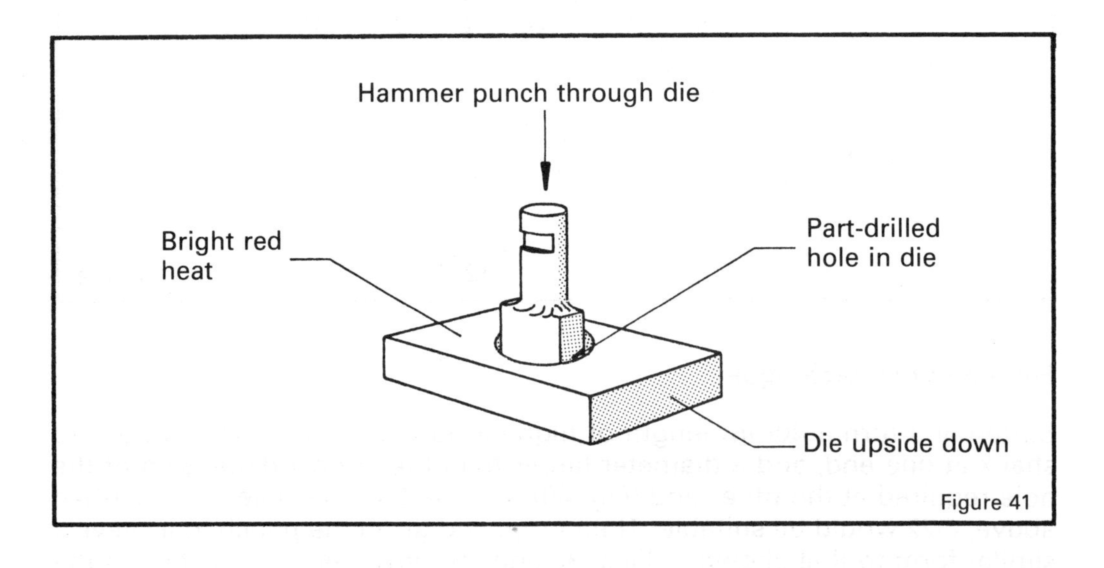

Figure 41

Let the die cool slowly. Check the cutting edges of the punch for hardness with a file. If they have softened, re-harden them. When the die has cooled, check that the punch fits into the die when presented from the top side. The part-drilled hole is the lower part of the die. File the die, if necessary, to achieve a close fit for the punch. If there is a large gap, the die can be closed up by heating to a bright red and hammering near to, but **not** at the edge of, the die cavity with a centrepunch or chisel (Fig. 42). This will close up the hole slightly. It can then be filed out if necessary to fit the punch exactly.

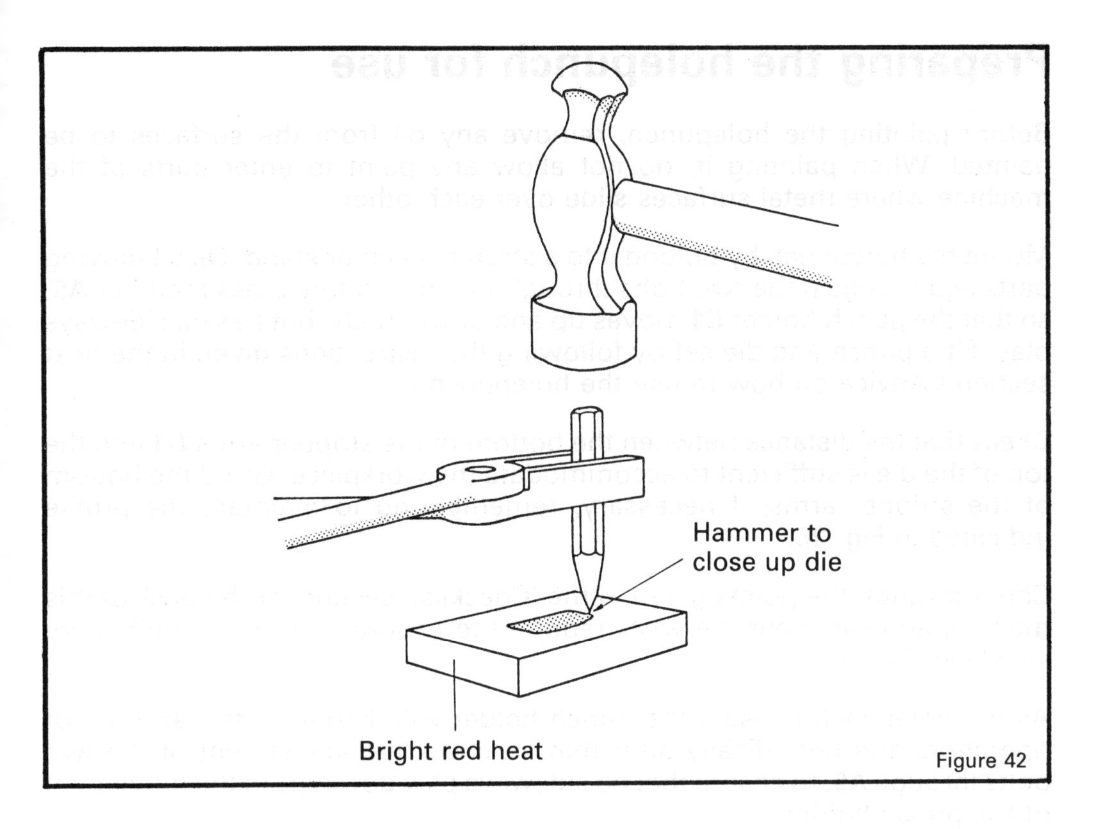

Figure 42

Put the punch in the punch holder and fit the die mounting plates **F1** and **F2** in the frame. Position the die **F3** and weld it to **F1** and **F2**, as indicated in fig. 36. After the welds have cooled, check that the die assembly can be mounted either way round. File **F1** and **F2** if necessary, to allow this. Re-harden the die by heating to a bright red and quenching in oil, as described previously.

Finally, check that both counterweights **C4** hit the punch holder operating pin **C9** at the same time when the handle **C8** is brought fully down. If not, grind a hollow in one of the counterweights so that they do. This is most important, as otherwise the punch is twisted slightly at the bottom of its travel. If the punch is rectangular, great stress can be placed on various components, causing them to crack.

Preparing the holepunch for use

Before painting the holepunch, remove any oil from the surfaces to be painted. When painting it, do not allow any paint to enter parts of the machine where metal surfaces slide over each other.

Mount the holepunch by bolting it to a strong bench or stand. Oil all moving parts again. Adjust the two bolts through the front frame cross member **A5**, so that the punch holder **C1** moves up and down freely, but has no sideways play. Fit a punch and die set by following the instructions given in the next section ('Advice on how to use the holepunch').

Check that the distance between the bottom of the stripper arms **D1** and the top of the die is sufficient to accommodate the workpiece. Grind the bottom of the stripper arms, if necessary, remembering to maintain the profile indicated in Fig. 26.

Check through the points given in the 'Checklist' section, at the back of this manual, and carry out the work required to ensure that all the points are satisfactorily met.

As the holepunch is used, the punch holder will 'bed in'. After an hour of operation, and periodically after that, re-check the adjustment of the two bolts through **A5**, to ensure that no sideways play develops in the movement of the punch holder.

Advice on how to use the holepunch

Always fit the punch into the punch holder first, and tighten the handscrew. Then position the die, and make sure that the punch fits centrally into the die before tightening the handscrews in the die mounting assembly. Check the alignment again before cutting any metal. Failure to follow this procedure will result quickly in the breaking of the punch, especially with the smaller punch sizes.

Note that the depth gauge can be positioned on top of the die, close to the punch, or it can be turned over and placed further back, to locate workpieces that have an edge which is lower than the die.

Always punch the holes required before carrying out other forming operations. Once the workpiece has been folded, for example, it may not be possible to fit in onto the die for punching.

Checklist

1. Check that the handle operates smoothly over the full extent of its travel. It should have little sideways movement.

2. When a punch and die set is fitted, check that the punch fits centrally into the die, with an even and small gap round the punch.

3. With the handle fully down, check that the counterweights on both sides are touching the pin through the top of the punch holder.

4. The handle should stay in the upper position, without falling down.

5. The punch holder must have no sideways movement.

6. The handscrews in the punch holder, the die mounting and the depth gauge should all operate freely and easily.

7. The stripper arms should swing freely when hanging down. Their bottom surfaces should be ground parallel to the die surface.

8. The holepunch should be bolted to a sturdy bench or stand.